Solutions Manual

Elementary Differential Equations

WITH

Boundary Value Problems

Solutions Manual

Elementary Differential Equations

WITH

Boundary Value Problems

THIRD EDITION

C.H. Edwards, Jr. / David E. Penney
The University of Georgia

PRENTICE HALL, *Englewood Cliffs, New Jersey* 07632

Editorial /production supervision
and interior design: **Michael Hyde**
Acquisitions editor: **Steve Conmy**
Supplement acquisitions editor: **Mary Hornby**
Prepress buyer: **Paula Massenaro**
Manufacturing buyer: **Lori Bulwin**

Printed in the United States of America

10   9   8   7   6   5   4   3   2   1

0-13-253428-2

CONTENTS

Chapter 3: POWER SERIES SOLUTIONS OF LINEAR EQUATIONS

Chapter 4: THE LAPLACE TRANSFORM

Chapter 5: LINEAR SYSTEMS OF DIFFERENTIAL EQUATIONS

Chapter 6: NUMERICAL METHODS

Chapter 7: NONLINEAR DIFFERENTIAL EQUATIONS
AND SYSTEMS

Chapter 8: FOURIER SERIES AND SEPARATION OF VARIABLES

Chapter 9: EIGENVALUES AND BOUNDARY VALUE PROBLEMS

APPENDIX

PREFACE

This is a solutions manual to accompany the textbook *ELEMENTARY DIFFERENTIAL EQUATIONS with Boundary Value Problems* (1993) by C. H. Edwards, Jr. and David E. Penney. We provide here either the answer or a solution to essentially every problem in the text. In some cases only the answer is given to a problem that is entirely routine or involves only a standard elementary technique, but otherwise at least an outline of the principal steps in a complete solution is offered.

In every case our goal is to provide just that amount of detail in the way of answers, hints, and suggestions that will supplement the text discussion most instructively. In most cases we believe that giving the highlights to an argument or computation, but leaving something for the student to do, is better than including a complete solution that students could simply transcribe. Many of the sections in this manual begin with comments on the priority of topics within the corresponding section of the text.

The purpose of this manual is assistance in the teaching and learning of the subject of elementary differential equations. To this end we invite suggestions from those who use it (both professors and students) as to what features might be added or improved to increase its usefulness in future editions.

We are indebted to Alice F. Edwards for her assistance in preparing the camera-ready copy for this manual.

C. H. E., Jr.
D. E. P.

Solutions Manual

Elementary Differential Equations

WITH

Boundary Value Problems

CHAPTER 1

FIRST ORDER DIFFERENTIAL EQUATIONS

SECTION 1.1

INTRODUCTION

The main purpose of Section 1.1 is simply to introduce the basic notation and terminology of differential equations, and to tell the student what is meant by a *solution* of a differential equation. Also, the use of differential equations in the *mathematical modeling* of real-world phenomena is outlined.

Problems 1-12 are routine verifications by direct substitution of the indicated solutions into the given differential equations. As an example we include the solution of Problem 11.

11. If $y = y_1 = x^{-2}$ then

$$y' = -2x^{-3} \quad \text{and} \quad y'' = 6x^{-4},$$

so

$$x^2 y'' + 5xy' + 4y = x^2(6x^{-4}) + 5x(-2x^{-3}) + 4(x^{-2})$$

$$= 6x^{-2} - 10x^{-2} + 4x^{-2} = 0.$$

If $y = y_2 = x^{-2}\ln x$ then

$$y' = x^{-3} - 2x^{-3}\ln x \quad \text{and} \quad y'' = -5x^{-4} + 6x^{-4}\ln x,$$

so

$x^2y'' + 5xy' + 4y$

$$= x^2(-5x^{-4} + 6x^{-4}\ln x) + 5x(x^{-3} - 2x^{-3}\ln x) + 4(x^{-2}\ln x)$$

$$= (-5x^{-2} + 5x^{-2}) + (6x^{-2} - 10x^{-2} + 4x^{-2})\ln x = 0.$$

13. $r = 2/3$ 14. $r = \pm 1/2$

15. $r = -2,1$

16.   If $y = e^{rx}$ then $y' = re^{rx}$ and $y'' = r^2 e^{rx}$,
      so
      $$3y'' + 3y' - 4y = (3r^2 + 3r - 4)e^{rx} = 0.$$

The solutions of the quadratic equation $3r^2 + 3r - 4 = 0$
are $r = (-3 \pm \sqrt{57})/6$.

17. $C = 2$ 18. $C = 3$

19. $C = 6$ 20. $C = 11$

21. $C = 7$

22.   Substitution of $x = y = 0$ in $y(x) = \ln(x + C)$ yields
      $0 = \ln C$,   so $C = 1$.

23. $C = 56$ 24. $C = 17$

25. $C = \pi/4$

26.   Substitution of $x = \pi$ and $y = 0$ in $y(x) = (x + C)\cos x$
      yields $0 = (\pi + C)(-1)$,   so $C = -\pi$.

27. $y' = x + y$

28. $y' = (y - 0)/(x - x/2) = 2y/x$

29.   If m is the slope of the tangent line and m' the slope

of the normal line at (x,y), then $mm' = -1$ yields

$$m' = -1/y' = (y - 1)/(x - 0).$$

Therefore $y' = x/(1 - y)$.

30. $D_x(k + x^2) = 2x$, so the orthogonality relation $mm' = -1$ implies that $y = g(x)$ satisfies the differential equation $y' = -1/2x$.

31. $y' = (y - x)/(x + y)$ 32. $dP/dt = k\sqrt{P}$

33. $dv/dt = kv^2$ 34. $dv/dt = k(250 - v)$

35. $dN/dt = k(P - N)$ 36. $dN/dt = kN(P - N)$

37. $y = 1$ or $y = x$ 38. $y = e^x$

39. $y = x^2$ 40. $y = 1$ or $y = -1$

41. $y = e^x/2$ 42. $y = \cos x$ or $y = \sin x$

43. (a) $y(10)$ yields $10 = 1/(C - 10)$, so $C = 101/10$.

 (b) There is no such value of C, but the constant function $y(x) = 0$ satisfies the conditions $y' = y^2$ and $y(0) = 0$.

44. (b) Obviously the functions $u(x) = -x^4$ and $v(x) = +x^4$ satisfy $xy' = 4y$. But $u'(x) = -4x^3$ and $v'(x) = +4x^3$ match at $x = 0$, where both are zero. Hence the given piecewise-defined function $y(x)$ is differentiable, and therefore satisfies the differential equation because $u(x)$ and $v(x)$ do so, for $x \leq 0$ and $x \geq 0$, respectively.

SOLUTION BY DIRECT INTEGRATION

This section introduces *general solutions* and *particular solutions* in the very simplest situation -- a differential equation of the form $y' = f(x)$ -- where only direct integration and evaluation of the constant of integration are involved. Students should review carefully the elementary concepts of velocity and acceleration, as well as the fps and cgs unit systems.

1.   If $y' = 2x + 1$ then integration yields

$$y(x) = \int (2x + 1) \, dx = x^2 + x + C.$$

Then substitution of $x = 0$, $y = 3$ gives
$3 = 0 + 0 + C = C$, so

$$y(x) = x^2 + x + 3.$$

The solutions in Problems 2-9 follow the pattern of Problem 1, and only the answers are given.

2. $y = (x - 2)^4/4 + 1$ 3. $y = (2x^{3/2} - 16)/3$

4. $y = -1/x + 6$ 5. $y = 2(x + 2)^{1/2} - 5$

6. $y = [(x^2 + 9)^{3/2} - 125]/3$ 7. $y = 10 \tan^{-1}x$

8. $y = (1/2)\sin 2x + 1$ 9. $y = \sin^{-1}x$

10. $y = -(1 + x)e^{-x} + 2$

11.   If $a(t) = 50$ then

$$v = \int 50 \, dt = 50t + v_0 = 50t + 10.$$

Hence

$$x = \int (50t + 10) \, dt$$

$$= 25t^2 + 10t + x_0 = 25t^2 + 10t + 20.$$

12. $x = -10t^2 - 15t + 5$ 　　　　　　13. $x = t^3/2 + 5t$

14. $x = t^3/3 + t^2/2 - 7t + 4$

15. $x = (1/3)(t + 3)^4 - 37t - 26$

16. $x = 4(t + 4)^{3/2} - t^2/2 + 4$

17. $x = [1/(t + 1) + t - 1]/2$

18. $v = -32t$ and $y = -16t^2 + 400$, so the ball hits the ground $(y = 0)$ when $t = 5$ sec, and then $v = -32(5) = -160$ ft/sec.

19. $a = -10$ m/s^2 and $v_0 = 100$ km/h ≈ 27.78 m/s, so $v = -10t + 27.78$, and hence

$$x(t) = -5t^2 + 27.78t.$$

The car stops when $v = 0$, $t \approx 2.78$, and thus the distance traveled before stopping is $x(2.78) \approx 38.59$ meters.

20. $v = -32t + 160$, so the ball reaches its maximum height $(v = 0)$ when $t = 5$ sec. Now

$$y = -16t^2 + 160t,$$

so $y_{max} = y(5) = 400$ ft. It hits the ground $(y = 0)$ when $t = 10$ sec.

21. $a = -9.8 \text{ m/s}^2$ so $v = -9.8t - 10$ and

$$y = -4.9t^2 - 10t + y_0.$$

The ball hits the ground when $y = 0$ and

$$v = -9.8t - 10 = -60,$$

so $t \approx 5.10$ s. Hence

$$y_0 = 4.9(5.10)^2 + 10(5.10) \approx 178.57 \text{ m}.$$

22. $v = -32t - 40$ and $y = -16t^2 - 40t + 555$. The ball hits
     the ground ($y = 0$) when $t \approx 4.77$ sec, with velocity
     $v = v(4.77) \approx -192.64$ ft/sec, an impact speed of about 131
     mph.

23. The height of the bomb after t seconds is $y = -16t^2 +$
     800, so $y = 400$ when $t = 5$. Hence we want the
     projectile to attain a height of 400 feet in 3 seconds.
     The projectile's height after t seconds is

$$y = -16t^2 + v_0t,$$

and substitution of $t = 3$, $y = 400$ yields $v_0 = 544/3 \approx$
181.33 ft/sec.

24. Taking $x_0 = 0$ and $v_0 = 60$ mph $= 88$ ft/sec, we get

$$v = -at + 88,$$

and $v = 0$ yields $t = 88/a$. Substituting this value of t
and $x = 176$ in

$$x = -at^2/2 + 88t,$$

we solve for $a = 22$ ft/sec^2. Hence the car skids for $t =$
$88/22 = 4$ sec.

25. If $a = -50$ ft/sec^2 and $x_0 = 0$ then the car's velocity and position at time t are given by

$$v = -50t + v_0, \qquad x = -25t^2 + v_0 t.$$

It stops when $v = 0$ so $v_0 = 50t$, and hence when

$$x = 225 = -25t^2 + (50t)t = 25t^2.$$

Thus $t = 3$ sec so

$$v_0 = 50(3) = 150 \text{ ft/sec} \approx 102 \text{ mph}.$$

26. Starting with $x_0 = 0$ and $v_0 = 50$ km/h $= 5 \cdot 10^4$ m/h, we find by the method of Problem 24 that the car's deceleration is $a = (25/3) \cdot 10^7$ m/h^2. Then, starting with $x_0 = 0$ and $v_0 = 100$ km/h $= 10^5$ m/h, we substitute $t = v_0/a$ into

$$x = -at^2 + v_0 t$$

and find that $x = 60$ m when $v = 0$. Thus doubling the initial velocity quadruples the distance the car skids.

27. If $v_0 = 0$ and $y_0 = 20$ then

$$v = -at \quad \text{and} \quad y = -0.5at^2 + 20.$$

Substitution of $t = 2$, $y = 0$ yields $a = 10$ ft/sec^2. If $v_0 = 0$ and $y_0 = 200$ then

$$v = -10t \quad \text{and} \quad y = -5t^2 + 200.$$

Hence $y = 0$ when $t = \sqrt{40} = 2\sqrt{10}$ sec and $v = -20\sqrt{10} \approx 63.25$ ft/sec.

28. <u>On Earth</u>: $v = -32t + v_0$, so $t = v_0/32$ at maximum height (when $v = 0$). Substituting this value of t and $y = 144$ in

$$y = -16t^2 + v_0t,$$

we solve for $v_0 = 96$ ft/sec as the initial speed with which the person can throw a ball straight upward.

On Planet Gzyx: From Problem 27, the surface gravitational acceleration on planet Gzyx is $a = 10$ ft/sec^2, so

$$v = -10t + 96 \qquad \text{and} \qquad y = -5t^2 + 96t.$$

Therefore $v = 0$ yields $t = 9.6$ sec, and thence $y_{max} = y(9.6) = 460.8$ ft is the height a ball will reach if its initial velocity is 96 ft/sec.

29. If $v_0 = 0$ and $y_0 = h$ then the stone's velocity and height are given by

$$v = -gt, \qquad y = -0.5gt^2 + h.$$

Hence $y = 0$ when $t = \sqrt{(2h/g)}$ so

$$v = -g\sqrt{(2h/g)} = -\sqrt{(2gh)}.$$

30. The method of solution is precisely the same as that in Problem 28. We find first that, on Earth, the woman must jump straight upward with initial velocity $v_0 = 12$ ft/sec to reach a maximum height of 2.25 ft. Then we find that, on the Moon, this initial velocity yields a maximum height of about 13.58 ft. For a complete solution (except with maximum height 4 ft on Earth), see Example 3 in Section 4.9 of Edwards & Penney, *Calculus and Analytic Geometry* (3rd edition, Prentice-Hall 1990).

31. We use units of miles and hours. If $x_0 = v_0 = 0$ then the car's velocity and position after t hours are given by

$$v = at, \qquad x = 0.5at^2.$$

Since $v = 60$ when $t = 5/6$, the velocity equation yields $a = 72$ mi/hr^2. Hence the distance traveled by 12:50 pm is

$$x = (0.5)(72)(5/6)^2 = 25 \text{ miles.}$$

32. Again we have

$$v = at, \qquad x = 0.5at^2.$$

But now $v = 60$ when $x = 35$. Substitution of $a = 60/t$ (from the velocity equation) into the position equation yields

$$35 = (0.5)(60/t)(t^2) = 30t,$$

whence $t = 7/6$ hr, that is, 1:10 pm.

33. Integration of $y' = (9/v_S)(1 - 4x^2)$ yields

$$y = (3/v_S)(3x - 4x^3) + C,$$

and the initial condition $y(-1/2) = 0$ gives $C = 3/v_S$. Hence the swimmer's trajectory is

$$y(x) = (3/v_S)(3x - 4x^3 + 1).$$

Substitution of $y(1/2) = 1$ now gives $v_S = 6$ mph.

34. Integration of $y' = 3(1 - 16x^4)$ yields

$$y = 3x - (48/5)x^5 + C,$$

and the initial condition $y(-1/2) = 0$ gives $C = 6/5$. Hence the swimmer's trajectory is
$$y(x) = (1/5)(15x - 48x^5 + 6),$$
so his downstream drift is $y(1/2) = 2.4$ miles.

EXISTENCE AND UNIQUENESS OF SOLUTIONS

As pointed out in the text, the instructor may choose to delay covering Section 1.3 until later in Chapter 1. However, before proceeding to Chapter 2, it is important that students come to grips at some point with the question of the existence of a unique solution of a differential equation - - and realize that it makes no sense to look for *the* solution without knowing in advance that it exists. The instructor may prefer to combine existence and uniqueness by simplifying the statement of the existence-uniqueness theorem as follows:

> *Suppose the function* f(x,y) *and the partial derivative* ∂f/∂y *are continuous in some neighborhood of the point* (a,b). *Then the initial value problem*
>
> $$y' = f(x,y), \quad y(a) = b$$
>
> *has a unique solution in some neighborhood of the point* a.

Slope fields and geometrical solution curves are introduced in this section as a concrete aid in visualizing solutions and existence-uniqueness questions, but do not play a substantial role subsequently in the text. Solution curves corresponding to the slope fields in Problems 1-10 are included in the answers section of the text and will not be duplicated here.

11. Each isocline $x - 1 = C$ is a vertical straight line.

12. Each isocline $x + y = C$ is a straight line with slope $m = -1$.

13. Each isocline $y^2 = C \geq 0$, that is, $y = \sqrt{C}$ or $y = -\sqrt{C}$, is a horizontal straight line.

14. Each isocline $y^{1/3} = C$, that is, $y = C^3$, is a horizontal straight line.

15. Each isocline $y/x = C$, or $y = Cx$, is a straight line through the origin.

16. Each isocline $x^2 - y^2 = C$ is a hyperbola that opens along the x-axis if $C > 0$, along the y-axis if $C < 0$.

17. Each isocline $xy = C$ is a rectangular hyperbola that opens along the line $y = x$ if $C > 0$, along $y = -x$ if $C < 0$.

18. Each isocline $x - y^2 = C$, or $y^2 = x - C$, is a translated parabola that opens along the x-axis.

19. Each isocline $y - x^2 = C$, or $x^2 = y - C$, is a translated parabola that opens along the y-axis.

20. Each isocline is an exponential graph of the form $y = Ce^x$.

21. Because both $f(x,y) = 2x^2 y^2$ and $\partial f/\partial y = 4x^2 y$ are continuous everywhere, the existence-uniqueness theorem of Section 1.3 guarantees the existence of a unique solution in some neighborhood of $x = 1$.

22. Both $f(x,y) = x \ln y$ and $\partial f/\partial y = x/y$ are continuous in a neighborhood of $(1,1)$, so the theorem guarantees the existence of a unique solution in some neighborhood of $x = 1$.

23. Both $f(x,y) = y^{1/3}$ and $\partial f/\partial y = (1/3)y^{-2/3}$ are continuous near $(0,1)$, so the theorem guarantees the existence of a unique solution in some neighborhood of $x = 0$.

24. $f(x,y) = y^{1/3}$ is continuous in a neighborhood of $(0,0)$, but $\partial f/\partial y = (1/3)y^{-2/3}$ is not, so the theorem guarantees existence but not uniqueness in some neighborhood of $x = 0$.

25. $f(x,y) = (x - y)^{1/2}$ is not continuous at $(2,2)$ because it is not even defined if $y > x$. Hence the theorem guarantees neither existence nor uniqueness in any neighborhood of the point $x = 2$.

26. Both $f(x,y) = (x - y)^{1/2}$ and $\partial f/\partial y = -(1/2)(x - y)^{-1/2}$ are continuous in a neighborhood of $(2,1)$, so the theorem guarantees both existence and uniqueness of a solution in some neighborhood of $x = 2$.

27. Both $f(x,y) = x/y$ and $\partial f/\partial y = -x/y^2$ are continuous near $(0,1)$, so the theorem guarantees both existence and uniqueness of a solution in some neighborhood of $x = 0$.

28. Neither $f(x,y) = x/y$ nor $\partial f/\partial y = -x/y^2$ is continuous near $(1,0)$, so the existence-uniqueness theorem guarantees nothing.

29. Both $f(x,y) = \ln(1 + y^2)$ and $\partial f/\partial y = 2y/(1 + y^2)$ are continuous near $(0,0)$, so the theorem guarantees the existence of a unique solution near $x = 0$.

30. Both $f(x,y) = x^2 - y^2$ and $\partial f/\partial y = -2y$ are continuous near $(0,1)$, so the theorem guarantees both existence and uniqueness of a solution in some neighborhood of $x = 0$.

31. If $f(x,y) = -(1 - y^2)^{1/2}$ then $\partial f/\partial y = y(1 - y^2)^{-1/2}$ is not continuous when $y = 1$, so the theorem does not guarantee uniqueness.

32. The two solutions are $y_1(x) = 0$ (constant) and $y_2(x) = x^3$.

35. The isoclines of $y' = y/x$ are the straight lines $y = Cx$ through the origin, and $y' = C$ at points of $y = Cx$, so it appears that these same straight lines are the solution curves of $xy' = y$. Then we observe that there is

    (i)   a unique one of these lines through any point not on the y-axis;

    (ii) no such line through any point on the y-axis other than the origin; and

    (iii) infinitely many such lines through the origin.

36. $f(x,y) = 4xy^{1/2}$ and $\partial f/\partial y = 2xy^{-1/2}$ are continuous if $y > 0$, so for all a and all $b > 0$ there exists a unique solution near $x = a$ such that $y(a) = b$. If $b = 0$ then the theorem guarantees neither existence nor uniqueness. For any a, both $y_1(x) = 0$ and $y_2(x) = (x^2 - a^2)^2$ are solutions with $y(a) = 0$. Thus we have existence but not uniqueness near points on the x-axis.

SECTION 1.4

SEPARABLE EQUATIONS AND APPLICATIONS

Of course it should be emphasized to students that the possibility of separating the variables is the first one you look for. The general concept of natural growth and decay is important for all differential equations students, but the particular applications in this section are optional. Torricelli's law in the form of Equation (19) leads to some nice concrete examples and problems.

1. $y = C \exp(-x^2)$ 2. $y = 1/(x^2 + C)$

3. $y = Ce^{-\cos x}$ 4. $y = C(1 + x)^4$

5. $y = \sin(\sqrt{x} + C)$ 6. $y = (x^{3/2} + C)^2$

7. $y = (2x^{4/3} + C)^{3/2}$ 8. $y = \sin^{-1}(x^2 + C)$

9.   The partial fraction decomposition

$$2/(1 - x^2) = 1/(1 + x) + 1/(1 - x)$$

leads to $y = C(1 + x)/(1 - x)$.

10. $y = [x - C(1 + x)]/[1 + C(1 + x)]$

11. $y = (C - x^2)^{-1/2}$ 12. $y^2 + 1 = C \exp(x^2)$

13. $(1/4)\ln(y^4 + 1) = \sin x + C$

14. $3y + 2y^{3/2} = 3x + 2x^{3/2} + C$

15. $1/3y^3 - 2/y = 1/x + \ln|x| + C$

16. $\sec y = C(x^2 + 1)^{1/2}$

17. $\ln|1 + y| = x + (1/2)x^2 + C$

18. $y = \tan(C - x - 1/x)$

19. $y = 2 \exp(e^x)$

20. $y = \tan(x^3 + \pi/4)$

21. $y^2 = 1 + (x^2 - 16)^{1/2}$

22. $|y| = 3 \exp(x^4 - x)$

23. $\ln(2y - 1) = 2(x - 1)$

24. $y = (\pi/2)\sin x$

25. $\ln y = x^2 - 1 + \ln x$

26. $y = 1/(1 - x^2 - x^3)$

27. About 51,840 persons   28. About 3.87 hours

29. About 14,735 years

30. The purported relic is about 686 years old.

31. $21,103.48   32. $44.52

33. 2585 mg   34. About 35 years

35. Taking t = 0 when the body was formed and t = T now, the
     amount Q(t) of lead in the body at time t (in years) is
     given by $Q(t) = Q_0 e^{-kt}$, where $k = (\ln 2)/(4.51 \cdot 10^9)$. The
     given information tells us that

$$\frac{Q(T)}{Q_0 - Q(T)} = 0.9.$$

     After substituting $Q(T) = Q_0 e^{-kt}$, we solve readily for e^{kt}
     = 19/9, so

$$T = (1/k)\ln(19/9) \approx 4.86 \cdot 10^9.$$

     Thus the body was formed approximately 4.86 billion years
     ago.

36. Taking t = 0 when the rock contained only potassium and
     t = T now, the amount Q(t) of potassium in the rock at
     time t (in years) is given by $Q(t) = Q_0 e^{-kt}$, where $k =$
     $(\ln 2)/(1.28 \cdot 10^9)$. The given information tells us that the
     amount A(t) of argon at time t is

$$A(t) = \frac{1}{9}\left[Q_0 - Q(t)\right]$$

     and also that A(T) = Q(T). Thus

$$Q_0 - Q(t) = 9Q(T).$$

     After substituting $Q(T) = Q_0 e^{-kt}$, we solve readily for

$$T = (\ln 10/\ln 2)(1.28 \cdot 10^9) \approx 4.25 \cdot 10^9.$$

Thus the age of the rock is about 4.25 billion years.

37. Because $A = 0$ the differential equation reduces to $T' = kT$, so $T(t) = 25e^{-kt}$. The fact that $T(20) = 15$ yields $k = (1/20)\ln(5/3)$, and finally we solve

$$5 = 25e^{-kt} \quad \text{for} \quad t = (\ln 5)/k \approx 63 \text{ min.}$$

38. About 2.41 minutes

39. (a) 0.495 m   (b) $(8.32 \times 10^{-7})I_0$ (c) 3.29 m

40. (a) 20.486 inches; 9.604 inches
    (b) 3.45 miles, or about 18,200 feet

41. (a) $A' = rA + Q$
    (b) The solution of the differential equation with $A(0) = 0$ is given by

    $$rA + Q = Qe^{rt}.$$

    When we substitute $A = 40$ (thousand), $r = 0.11$, and $t = 18$, we find that $Q = 0.70482$, that is, \$704.82 per year.

42. The answer is about 5.99 billion years. For a complete solution see Example 3 in Section 7.5 of Edwards & Penney, *Calculus and Analytic Geometry* (3rd edition, Prentice-Hall 1990).

43. After 66 min 40 sec; this problem is just like Example 6 in the text.

44. (b) By separating the variables we solve the differential equation for

$$c - rP(t) = (c - rP_0)e^{rt}.$$

With $P(t) = 0$ this yields

$$c = rP_0e^{rt}/(e^{rt} - 1).$$

With $P_0 = 10,800$, $t = 60$, and $r = 0.010$ we get $239.37 for the monthly payment at 12% annual interest. With $r = 0.015$ we get $272.99 for the monthly payment at 18% annual interest.

45. If $N(t)$ denotes the number of people (in thousands) who have heard the rumor after t days, then the initial value problem is

$$N' = k(100 - N), \qquad N(0) = 0$$

and we are given that $N(7) = 10$. The answer is $t \approx 46$ days.

46. With $A(y) = $ constant, Equation (19) in the text takes the form

$$\frac{dy}{dt} = k\sqrt{y}.$$

We readily solve this equation for

$$2\sqrt{y} = -kt + C.$$

The condition $y(0) = 9$ yields $C = 6$, and then $y(1) = 4$ yields $k = 2$. Thus the depth at time t (in hours) is $y(t) = (3 - t)^2$, and hence it takes 3 hours for the tank to empty.

47. With $A = \pi(3)^2$ and $a = \pi(1/12)^2$, and taking $g = 32$ ft/sec^2, Equation (20) reduces to $162y' = -\sqrt{y}$. The solution such that $y = 9$ when $t = 0$ is given by

$$324\sqrt{y} = -t + 972.$$

Hence $y = 0$ when $t = 972$ sec $= 16$ min 12 sec.

48. The radius of the cross-section of the cone at height y is proportional to y, so $A(y)$ is proportional to y^2. Therefore Equation (20) takes the form

$$y^2 y' = - k\sqrt{y},$$

and a general solution is given by

$$2y^{5/2} = - 5kt + C.$$

The initial condition $y(0) = 16$ yields $C = 2048$, and then $y(1) = 9$ implies that $5k = 1562$. Hence $y = 0$ when

$$t = C/5k = 2048/1562 \approx 1.31 \text{ hr.}$$

49. The solution of $y' = - k\sqrt{y}$ is given by

$$2\sqrt{y} = - kt + C.$$

The initial condition $y(0) = h$ (the height of the cylinder) yields $C = 2\sqrt{h}$. Then substitution of $t = T$, $y = 0$ gives $k = (2\sqrt{h})/T$. It follows that

$$y = h(1 - t/T)^2.$$

If r denotes the radius of the cylinder, then

$$V(y) = \pi r^2 y = \pi r^2 h(1 - t/T)^2 = V_0(1 - t/T)^2.$$

50. Since $x = y^{3/4}$, the cross-sectional area is $A(y) = \pi x^2 = \pi y^{3/2}$. Hence the general equation $A(y)y' = -a\sqrt{(2gy)}$ reduces to the differential equation $yy' = -k$ with general solution

$$(1/2)y^2 = -kt + C.$$

The initial condition $y(0) = 12$ gives $C = 72$, and then $y(1) = 6$ yields $k = 54$. It follows that the depth at time t is

$$y(t) = \sqrt{(144 - 108t)},$$

so the tank is empty after $t = 4/3$ hr, that is, at 1:20 pm.

51. (a) Since $x^2 = by$, the cross-sectional area is $A(y) = \pi x^2 = \pi by^2$. Hence the equation $A(y)y' = -a\sqrt{(2gy)}$ reduces to the differential equation

$$y^{1/2}y' = -k = -(a/\pi b)\sqrt{(2g)}$$

with the general solution

$$(2/3)y^{3/2} = -kt + C.$$

The initial condition $y(0) = 4$ gives $C = 16/3$, and then $y(1) = 1$ yields $k = 14/3$. It follows that the depth at time t is

$$y(t) = (8 - 7t)^{2/3}.$$

(b) The tank is empty after $t = 8/7$ hr. that is, at 1:08:34 pm.

(c) We see above that $k = (a/\pi b)\sqrt{(2g)} = 14/3$. Substitution of $a = \pi r^2$, $b = 1$, $g = (32)(3600)^2$ ft/hr^2 yields $r = (1/60)\sqrt{(7/12)}$ ft ≈ 0.15 in for the radius of the bottom-hole.

52. With $g = 32$ ft/sec^2 and $a = \pi(1/12)^2$, Equation (19) simplifies to

$$A(y)\frac{dy}{dt} = -\frac{\pi}{18}\sqrt{y}.$$

If z denotes the distance from the center of the cylinder down to the fluid surface, then $y = 3 - z$ and $A(y) = 10(9 - z^2)^{1/2}$. Hence the equation above becomes

$$10(9 - z^2)^{1/2} \frac{dz}{dt} = \frac{\pi}{18} (3 - z)^{1/2},$$

$$180(3 + z)^{1/2} dz = \pi dt,$$

and integration yields

$$120(3 + z)^{3/2} = \pi t + C.$$

Now $z = 0$ when $t = 0$, so $C = 120(3)^{3/2}$. The tank is empty when $z = 3$ (that is, when $y = 0$) and thus after

$$t = (120/\pi)(6^{3/2} - 3^{3/2}) \approx 362.90 \text{ sec.}$$

It therefore takes about 6 min 3 sec for the fluid to drain completely.

53. $A(y) = \pi(8y - y^2)$ as in Example 7 in the text, but now $a = \pi/144$ in Equation (20), so the initial value problem is

$$18(8y - y^2)y' = - \sqrt{y}, \quad y(0) = 8.$$

We seek the value of t when $y = 0$. The answer is $t \approx 869$ sec = 14 min 29 sec.

54. The given rate of fall of the water level is $dy/dt = -4$ in/hr = $- (1/10800)$ ft/sec. With $A = \pi x^2$ and $a = \pi r^2$, Equation (19) is

$$(\pi x^2)(- 1/10800) = - (\pi r^2)\sqrt{(2gy)} = - 8\pi r^2 \sqrt{y}.$$

Hence the curve is of the form $y = kx^4$, and in order that it pass through (1,4) we must have $k = 4$. Comparing $\sqrt{y} = 2x^2$ with the equation above, we see that

$$(8r^2)(10800) = 1/2,$$

so the radius of the bottom hole is

$$r = 1/240\sqrt{3} \text{ ft} \approx 1/35 \text{ in.}$$

55. Let $t = 0$ at the time of death. Then the solution of the initial value problem

$$T' = k(70 - T), \quad T(0) = 98.6$$

is

$$T(t) = 70 + 28.6e^{-kt}.$$

If $t = a$ at 12 noon, then we know that

$$T(t) = 70 = 28.6e^{-ka} = 80,$$

$$T(a+1) = 70 + 28.6e^{-k(a+1)} = 75.$$

Hence

$$28.6e^{-ka} = 10 \quad \text{and} \quad 28.6e^{-k}e^{-ka} = 5.$$

It follows that $e^{-k} = 1/2$, so $k = \ln 2$. Finally the first of the previous two equations yields

$$a = (\ln 2.86)/(\ln 2) \approx 1.516 \text{ hr} \approx 1 \text{ hr} \quad 31 \text{ min,}$$

so the death occurred at 10:29 am.

56. Let $t = 0$ when it began to snow, and $t = t_0$ at 7:00 am. Let x denote distance along the road, with $x = 0$ where the snowplow begins at 7:00 am. If $y = ct$ is the snow depth at time t, w is the width of the road, and $v = dx/dt$ is the plow's velocity, then "plowing at a constant rate" means that $wyv = \text{constant}$. Hence our differential equation is of the form

$$k\frac{dx}{dt} = \frac{1}{t} .$$

The solution with $x = 0$ when $t = t_0$ is

$$t = t_0 e^{kx}.$$

We are given that $x = 2$ when $t = t_0 + 1$ and $x = 4$ when $t = t_0 + 3$, so it follows that

$$t_0 + 1 = t_0 e^{2k} \quad \text{and} \quad t_0 + 3 = t_0 e^{4k}.$$

Elimination of t_0 yields the equation

$$e^{4k} - 3e^{2k} + 2 = (e^{2k} - 1)(e^{2k} - 2) = 0,$$

so it follows (since $k > 0$) that $e^{2k} = 2$. Hence $t_0 + 1 = 2t_0$, so $t_0 = 1$. Thus it began to snow at 6 am.

57. We still have $t = t_0 e^{kx}$, but now the given information yields the conditions

$$t_0 + 1 = t_0 e^{4k} \quad \text{and} \quad t_0 + 2 = t_0 e^{7k}$$

at 8 am and 9 am, respectively. Elimination of t_0 gives the equation

$$2e^{4k} - e^{7k} - 1 = 0,$$

which we solve numerically for $k = 0.08276$. Using this value, we finally solve one of the preceding pair of equations for $t_0 = 2.5483$ hr $\approx$ 2 hr 33 min. Thus it began to snow at 4:27 am.

SECTION 1.5

LINEAR FIRST ORDER EQUATIONS

In addition to learning to recognize and solve linear first order equations, it is good preparation for Chapter 2 (higher order linear equations) for students to study carefully the existence-uniqueness theorem for first order linear equations in this section. Although we recommend that students *not* memorize Equation (10), it's hard to deny that many students can set up mixture problems most easily by substituting in the equation

$$\frac{dx}{dt} = r_i c_i - r_o c_o \, ,$$

where r_i and r_o are the incoming and outgoing flow rates, and c_i and $c_o = x/V$ are the incoming and outgoing concentrations. In each of Problems 1-25 we give the integrating factor $\rho(x) = \exp [\int P(x) dx]$ as well as the desired general or particular solution. It is important to write the differential equation precisely in the form $y' + P(x)y = Q(x)$ before attempting to calculate $\rho(x)$.

1. $\rho = e^x;\quad y = 2(1 - e^{-x})$

2. $\rho = e^{-2x};\quad y = 3xe^{2x}$

3. $\rho = e^{3x};\quad y = e^{-3x}(x^2 + C)$

4. $\rho = \exp(-x^2);\quad y = (x + C)\exp(-x^2)$

5. $\rho = x^2;\quad y = x + 4/x^2$

6. $\rho = x^5;\quad y = x^2 + 32/x^5$

7. $\rho = x^{1/2};\quad y = 5x^{1/2} + Cx^{-1/2}$

8. $\rho = x^{1/3};\quad y = 3x + Cx^{-1/3}$

9. $\rho = 1/x;\quad y = x(7 + \ln x)$

10. $\rho = x^{-3/2};\quad y = 3x^3 + Cx^{3/2}$

11. $\rho = xe^{-3x}$;   the general solution is $y(x) = Cx^{-1}e^{3x}$.   The particular solution such that $y(1) = 0$ is the constant function $y(x) = 0$ (obtained with $C = 0$).

12. $\rho = x^3$; $y = x^5/4 - 56/x^3$

13. $\rho = e^x$; $y = (e^x + e^{-x})/2$

14. $\rho = x^{-3}$; $y = x^3(10 + \ln x)$

15. $\rho = \exp(x^2)$; $y = [1 - 5\exp(-x^2)]/2$

16. $\rho = \exp(\sin x)$; $y = 1 + \exp(-\sin x)$

17. $\rho = 1 + x$; $y = (1 + \sin x)/(1 + x)$

18. $\rho = x^2$; $y = x^2(C + \sin x)^{\wedge}0$

19. $\rho = \sin x$; $y = (1/2)\sin x + C \csc x$

20. $\rho = \exp(-x - x^2/2)$; $y = \exp(x + x^2/2) - 1$

21. $\rho = x^{-3}$; $y = x^3 \sin x$

22. $\rho = \exp(-x^2)$; $y = (x^3 + 5)\exp(x^2)$

23. $\rho = x^{-3}e^{2x}$; $y = x^3(2 + Ce^{-2x})$

24. $\rho = (x^2 + 4)^{3/2}$; $3y = 1 + 16(x^2 + 4)^{-3/2}$

25. First we calculate

$$\int \frac{3x^3 \, dx}{x^2 + 1} = \int \left[3x - \frac{3x}{x^2 + 1} \right] dx$$

$$= (3/2)x^2 - (3/2)\ln(x^2 + 1).$$

It follows that

$$\rho = (x^2 + 1)^{-3/2}\exp(3x^2/2).$$

The desired particular solution is

$$y = [\exp(-3x^2/2)][3(x^2 + 1)^{3/2} - 2].$$

26. With $x' = dx/dy$, the differential equation is

$$y^3x' + 4y^2x = 1.$$

With y as the independent variable we find that the integrating factor is $\rho(y) = y^4$; the general solution is $x = 1/2y^2 + C/y^4$

27. $x' - x = ye^y$

$\rho(y) = e^{-y}$; $x = e^y(C + y^2/2)$

28. $(1 + y^2)x' - 2yx = 1$

$\rho(y) = (1 + y^2)^{-1}$

$2x = y + (1 + y^2)(C + \tan^{-1}y)$

29. $y = [\exp(x^2)][C + (\sqrt{\pi}/2)\operatorname{erf}(x)]$

30. After division of the given equation by 2x, multiplication by the integrating factor $\rho = x^{-1/2}$ yields

$$x^{-1/2}y' - (1/2)x^{-3/2}y = x^{-1/2}\cos x,$$

$$D_x(x^{-1/2}y) = x^{-1/2}\cos x,$$

$$x^{-1/2}y = C + \int_1^x t^{-1/2}\cos t \, dt$$

The initial condition $y(1) = 0$ implies that $C = 0$, so the desired particular solution is

$$y(x) = x^{1/2} \int_1^x t^{-1/2} \cos t \, dt.$$

32. (a) $y_p(x) = \sin x - \cos x$
    (b) $y(x) = Ce^{-x} + \sin x - \cos x$
    (c) $y(x) = 2e^{-x} + \sin x - \cos x$

33. After about 7 min 41 sec

34. About 22.2 days. For a complete solution see Example 4 in Section 7.6 of Edwards and Penney, *Calculus and Analytic Geometry* (3rd edition, Prentice-Hall 1990).

35. After about 10.85 years

36. (a) The volume of brine in the tank after t min is $V(t) = 60 - t$ gal, so the initial value problem is

    $$x'(t) = 2 - 3x/(60 - t), \quad x(0) = 0.$$

    The solution is

    $$x(t) = (60 - t) - (60 - t)^3/3600.$$

    (b) The maximum amount of salt in the tank is $40/\sqrt{3} \approx 23.09$ lb after $t = 60 - 20\sqrt{3} \approx 25.36$ min.

37. The volume of brine in the tank after t min is $V(t) = 100 + 2t$ gal, so the initial value problem is

    $$x'(t) = 5 - 3x/(100 + 2t), \quad x(0) = 50.$$

    The integrating factor $\rho(t) = (100 + 2t)^{3/2}$ leads to the solution

$$x(t) = 100 + 2t - 50{,}000(100 + 2t)^{-3/2}$$

such that $x(0) = 50$. The tank is full after $t = 150$ min, at which time $x(150) = 393.75$ lb.

38. (a) $x(t) = 50\,e^{-t/20}$

    (b) The solution of

$$\frac{dy}{dt} = \frac{5x}{100} - \frac{5y}{200} = \frac{5}{2}\,e^{-t/20} - \frac{y}{40}$$

with $y(0) = 50$ is

$$y(t) = 150\,e^{-t/40} - 100\,e^{-t/20}$$

    (c) $y_{max} = 56.25$ lb when $t = 40\,\ln(4/3) \approx 11.51$ min.

39. (a) The initial value problem

$$x' = -x/10, \qquad x(0) = 100$$

for Tank 1 has solution $x(t) = 100e^{-t/10}$. Then the initial value problem

$$y' = x/10 - y/10$$

$$= 10e^{-t/10} - y/10, \qquad y(0) = 0$$

for Tank 2 has solution $y(t) = 10te^{-t/10}$.

    (b) $y_{max} = y(10) = 100e^{-1} \approx 36.79$ gal.

40. (b) Assuming inductively that $x_n = t^n e^{-t/2}/n!\,2^n$, the equation for x_{n+1} is

$$x'_{n+1} = \frac{1}{2}x_n - \frac{1}{2}x_{n+1} = \frac{t^n\,e^{-t/2}}{n!\,2^{n+1}} - \frac{1}{2}x_{n+1}\,.$$

We easily solve this first order equation with $x_{n+1}(0) = 0$ and find that

$$x_{n+1} = t^{n+1}e^{-t/2}/(n+1)!2^{n+1},$$

thereby completing the proof by induction.

41. (a) $A'(t) = 0.06A + 0.12S = 0.06A + 3.6e^{0.05t}$

(b) The solution with $A(0) = 0$ is

$$A(t) = 360(e^{0.06t} - e^{0.05t}),$$

so $A(40) \approx 1308.283$ thousand dollars.

42. The mass of the hailstone at time t is $m = (4/3)\pi r^3 = (4/3)\pi k^3 t^3$. Then the equation $d(mv)/dt = mg$ simplifies to

$$tv' + 3v = gt.$$

The solution satisfying the initial condition $v(0) = 0$ is $v(t) = gt/4$, so $v'(t) = g/4$.

SECTION 1.6

SUBSTITUTION METHODS

It is traditional for every elementary differential equations text to include the particular types of equations that are found in this section. However, no one of them is especially important solely in its own right. Their real purpose (at this point in the course) is to familiarize students with the technique of transforming a differential equation by substitution. The subsection on airplane flight trajectories (together with Problems 44-47) is optional material and may be omitted if the instructor desires.

The differential equations in Problems 1-15 are homogeneous, so we make the substitutions

$$v = \frac{y}{x}, \qquad y = vx, \qquad \frac{dy}{dx} = v + x\frac{dv}{dx}.$$

1. $x^2 - 2xy - y^2 = C$ 2. $y^2 = x^2(\ln x + C)$

3. $y = x(C + \ln|x|)^2$

4. $\tan^{-1}(y/x) + (1/2)\ln(1 + y^2/x^2) = \ln x + C$

5. $\ln|xy| = C + x/y$ 6. $x = 2y(\ln y + C)$

7. $y^3 = 3x^3(C + \ln|x|)$ 8. $y = -x\ln(C - \ln x)$

9. $y = x/(C - \ln|x|)$ 10. $x^2 + 2y^2 = Cx^6$

11. $y = C(x^2 + y^2)$ 12. $4x^2 + y^2 = x^2(\ln x + C)^2$

13. $y + (x^2 + y^2)^{1/2} = Cx^2$

14. If $x > 0$ then the substitution $y = vx$ leads to

$$\ln x = \int \frac{v\, dv}{(1+v^2)^{1/2} - (1+v^2)}$$

$$= \frac{1}{2}\int \frac{du}{u^{1/2}(1 - u^{1/2})} \qquad (u = 1 + v^2)$$

$$= -\int \frac{dw}{w} = -\ln w + \ln C$$

with $w = 1 - u^{1/2}$. Back-substitution and simplification finally yields the implicit solution

$$x - (x^2 + y^2)^{1/2} = C.$$

15. $x^2(2xy + y^2) = C$

16. The substitution $v = x + y + 1$ leads to

$$x = \int \frac{dv}{1 + v^{1/2}} = \int \frac{2u\ du}{1 + u} \qquad (v = u^2)$$

$$= 2u - 2\ln(1 + u) + C$$

$$x = 2(x+y+1)^{1/2} - 2\ln[1 + (x+y+1)^{1/2}] + C$$

17. The substitution $v = 4x + y$ yields

$$y = 2\tan(2x + C) - 4x$$

18. The substitution $v = x + y$ leads to

$$y = \ln(x + y + 1) + C.$$

Problems 19-25 are Bernoulli equations. We indicate for each the appropriate substitution as specified in Equation (10) of this section.

19. $y^2 = x/(2 + Cx^5)$; $v = y^{-2}$

20. $y = [3 + C\exp(-3x^2)]^{1/3}$; $v = y^3$

21. $y^2 = 1/(Ce^{-2x} - 1)$; $v = y^{-2}$

22. $y = (Cx^6 + 15/7x)^{-1/3}$; $v = y^{-3}$

23. $y = (x + Cx^2)^{-3}$; $v = y^{-1/3}$

24. $y^2 = e^{2x}/(C + \ln x)$; $v = y^{-2}$

25. $2x^3y^3 = 3(1 + x^4)^{1/2} + C$; $v = y^3$

26. The substitution $v = y^3$ yields the linear equation $v' + v = e^{-x}$.

Solution: $y^3 = e^{-x}(x + C)$

27.   The substitution $v = y^3$ yields the linear equation
$xv' - v = 3x^4$.

Solution: $y = (x^4 + Cx)^{1/3}$

28.   The substitution $v = e^y$ yields the linear equation
$xv' - 2v = 2x^3e^{2x}$.

Solution: $y = \ln(Cx^2 + x^2e^{2x})$

29.   The substitution $v = \sin y$ yields the homogeneous equation
$2xvv' = 4x^2 + 3v^2$.

Solution: $\sin^2 y = Cx^3 - 4x^2$

30.   First we multiply each side of the given equation by e^y.
Then the substitution $v = e^y$ gives the homogeneous
equation $(x + v)v' = x - v$ of Problem 1 above.

Solution: $x^2 - 2xe^y - e^{2y} = C$

34.   The substitution $v = \ln y$, $y = e^v$, $y' = e^v v'$ yields the
linear equation $xv' + 2v = 4x^2$.

Solution: $y = \exp(x^2 + C/x^2)$

35.   Substitution: $x = u - 1$, $y = v - 2$

Solution: $x^2 - 2xy - y^2 - 2x - 6y = C$

36.   The substitution $x = u + 3$, $y = v - 2$ yields the
homogeneous equation

$$\frac{dv}{du} = \frac{-u + 2v}{4u - 3v}.$$

The substitution $v = pu$ leads to

$$\ln u = \int \frac{(4 - 3p)\,dp}{(3p + 1)(p - 1)}$$

Evaluation of this integral by partial fractions finally yields the implicit solution

$$(x + 3y + 3)^5 = C(- x + y + 5).$$

37. The substitution $v = x - y$ yields the separable equation $v' = 1 - \sin v$. With the aid of the identity

$$1/(1 - \sin v) = (1 + \sin v)/\cos^2 v$$

$$= \sec^2 v + \sec v \tan v$$

we obtain the solution

$$x = \tan(x - y) + \sec(x - y) + C.$$

40. $y = x + (2/\sqrt{\pi})[\exp(-x^2)][C + \operatorname{erf}(x)]^{-1}$

in terms of the error function $\operatorname{erf}(x)$ defined in Problem 29 of Section 1.5.

41. $y = x + (C - x)^{-1}$

42. The substitution $y' = C$ in the Clairaut equation immediately yields the general solution $y = Cx + g(C)$.

43. Clearly the line $y = Cx - C^2/4$ and the tangent line at $(C/2, C^2/4)$ to the parabola $y = x^2$ both have slope C.

45. With $a = 100$ and $k = 1/10$, Equation (19) is

$$y = 50[(x/100)^{9/10} - (x/100)^{11/10}].$$

The equation $y'(x) = 0$ then yields

$$(x/100)^{1/10} = (9/11)^{1/2},$$

so it follows that

$$y_{max} = 50[(9/11)^{9/2} - (9/11)^{11/2}] \approx 3.68 \text{ mi.}$$

47. (a) $y = 50[(x/100)^{1/2} - (x/100)^{3/2}]; \quad y(0) = 0$

(b) $y = 50[1 - (x/100)^2]; \quad y(0) = 50$

(c) $y = 50[(x/100)^{-1/2} - (x/100)^{5/2}];$

$y \longrightarrow \infty$ as $x \longrightarrow 0$

SECTION 1.7

EXACT EQUATIONS AND INTEGRATING FACTORS

The significance of this section is that it unifies and subsumes all of our first chapter methods for solving first order differential equations. Given such an equation to solve, we attempt to transform it -- by separating the variables, by multiplication by an integrating factor, by making a preliminary substitution, or whatever -- so that it takes the form

$$\frac{\partial F}{\partial x} \, dx + \frac{\partial F}{\partial y} \, dy = 0 ,$$

where the function $F(x,y)$ is evident or can be found. Then a general solution of our differential equation is defined implicitly by the equation $F(x,y) = C$.

The differential equations in Problems 1-12 are exact, and we give only the solutions that are found using the method of Example 2 in the text.

1. $x^2 + 3xy + y^2 = C$ 2. $2x^2 - xy + 3y^2 = C$

3. $x^3 + 2xy^2 + 2y^3 = C$ 4. $x^3 + x^2y^2 + y^4 = C$

5. $(1/4)x^4 + (1/3)y^3 + y \ln x = C$

6. $x + y^2 + e^{xy} = C$

7. $\sin x + x \ln y + e^y = C$

8. $x^2 + 2x \tan^{-1}y + \ln(1 + y^2) = C$

9. $x^3y^3 + xy^4 + (1/5)y^5 = C$

10. $e^x\sin y + x \tan y = C$

11. $x^2/y + y^2/x^3 + 2y^{1/2}$

12. $xy^{-2/3} + x^{-3/2}y = C$

13. $(y\ dx - x\ dy)/y^2 = y\ dy;\quad 2x = y^3 + Cy$

14. $(x\ dy - y\ dx)/x^2 = e^x dx\ ;\ y/x = e^x + C$

15. $(y\ dx + x\ dy) - dx - 2dy = 0;$
       $xy - x - 2y = C$

16. $(xy^2\ dx + x^2y\ dy) - (x\ dx + y\ dy) = 0$
       $x^2y^2 - (x^2 + y^2) = C$

17. $(3x^2y\ dx + x^3\ dy) + e^y\ dy = 0;\quad x^3y + e^y = C$

18. $(x^2 + y^2)dx - (x^2 + y^2)dy + (2x\ dx + 2y\ dy) = 0$

$$dx - dy + (2x\ dx + 2y\ dy)/(x^2 + y^2) = 0$$

$$x - y + \ln(x^2 + y^2) = C$$

19. $dx + (x\ dy - y\ dx)/(x^2 + y^2) = 0$

$$x + \tan^{-1}(y/x) = C$$

20. $(y^2\ dx + 2xy\ dy) + (e^x\sin y\ dx + e^x\cos y\ dy) = 0$

$$xy^2 + e^x\sin y = C$$

21. Integrating factor: $\rho(x) = x^3$; solution: $x^4y = C$

22. Integrating factor: $\rho(x) = x^2$; solution: $x^4 + x^3y^3 = C$

23. Integrating factor: $\rho(y) = y^{-2}$; solution: $x^2 + y^2 = Cy$

24. Integrating factor: $\rho(x) = x^2$
    Solution: $4x^5/5 + x^3\cos y = C$

25. Integrating factor: $\rho(y) = y^{-1}$
    Solution: $e^x + x \ln y + \sin y = C$

26. Integrating factor: $\rho(y) = 1/y^2$
    Solution: $2x^2 + 3y^2 + x/y = C$

27. Integrating factor: $\rho(y) = \sin y$
    Solution: $x^2 \sin y = C$

28. $(N_x - M_y)/M = - (\sec^2 y)/\tan y$, so the integrating factor is

$$\rho(y) = \exp(-\ln \tan y) = \cot y.$$

Multiplication by this integrating factor simply separates the variables.

Solution : $x + \ln(x \tan y) = C$.

29. Integrating factor: $\rho(x,y) = x^2y$

Solution: $x^7y^2 - x^3y^9 = C$

30. First we multiply the given equation by x^my^n to obtain an equivalent equation $M\,dx + N\,dy = 0$. When we equate the coefficients of x^my^{n+1} and of $x^{m+1}y^n$ in M_y and N_x, we get the two equations

$$3(n + 2) = 5(m + 1) , \quad 5(n + 1) = 6(m + 2)$$

whose solution is $m = 26/7$, $n = 41/7$. Multiplication of the original equation by the integrating factor

$$\rho(x,y) = x^{26/7}y^{41/7}$$

yields the exact equation

$$(3x^{26/7}y^{55/7} + 10x^{33/7}y^{48/7})\,dx$$

$$+ (5x^{33/7}y^{48/7} + 12x^{40/7}y^{41/7})\,dy = 0.$$

The solution is

$$4x^{33/7}y^{55/7} + 11x^{40/7}y^{48/7} = C.$$

31. Division of the given equation by $(x^2 + y^2)$ yields the equation

$$(y\,dx - x\,dy)/(x^2 + y^2) + (x\,dx + y\,dy)/(x^2 + y^2)^{1/2} = 0.$$

Solution: $\tan^{-1}(x/y) + (x^2 + y^2)^{1/2} = C$

32. When we divide the given equation by $(x^2 + y^2)$ and group terms, we get the equation

$$(y\,dx - x\,dy)/(x^2 + y^2) + (x\,dx + y\,dy)/(x^2 + y^2)^{-2} = 0.$$

The solution is

$$(x^2 + y^2)^{-1} - 2 \tan^{-1}(x/y) = C.$$

36. (a)   If x(t) denotes the length of rope overhanging after t
    sec, then its weight is ρxg,   and its average distance
    below the tabletop (x = 0, PE = 0) is x/2 .   Then
    conservation of total (kinetic plus potential) energy yields

$$KE + PE = (1/2)\rho L(x')^2 - (1/2)\rho gx^2 = C.$$

Differentiation of this equation gives

$$\rho Lx'x'' - \rho gxx' = 0,$$

which simplifies to

$$Lx'' = gx \; ; \quad \text{that is,} \quad x'' = 8x.$$

(b) By the method of Problem 38 in Section 1.6 we derive the
general solution

$$x(t) = A \cosh t\sqrt{8} + B \sinh t\sqrt{8} .$$

The intitial conditions x(0) = 1 and x'(0) = 0   imply
that   A = 1   and   B = 0,   so   x = cosh t$\sqrt{8}$.   Hence
x = L = 4   when

$$t = (1/\sqrt{8})\cosh^{-1}(4) \approx 0.7295 \text{ sec} .$$

SECTION 1.8

POPULATION MODELS

Sections 1.8 and 1.9 are optional. Unless an unusual amount of
time for applications is available, the instructor can choose
between the biological applications in this section and the
physical applications in the following section.

1. Given: $P' = -\delta P = -k\sqrt{P}$ and $P(0) = 900$, $P(6) = 441$
    Integration of the differential equation gives

    $$2\sqrt{P} = -kt + C,$$

    and then the given data yield $C = 60$ and $k = 3$.   Thus
    $2\sqrt{P} = -3t + 60$,   so $P = 0$ after $t = 20$ weeks.

2. (a) Starting with $dP/dt = k\sqrt{P}$,   we separate the variables
    and integrate to get

    $$P(t) = (kt/2 + C)^2.$$

    Clearly $P(0) = P_0$ implies $C = \sqrt{P_0}$.

    (b) If $P(t) = (kt/2 + 10)^2$,   then $P(6) = 169$ implies that
    $k = 1$.   Hence $P(t) = (t/2 + 10)^2$,   so there are   256   fish
    after   12   months.

3. (b) $P(t) \longrightarrow \infty$ as $t \longrightarrow 6$ (months)

4. Now, $dP/dt = -kP^2$ with $k > 0$,   and separation of
    variables yields

    $$P(t) = 1/(kt + C).$$

    Clearly $C = 1/P_0$,   so $P(t) = P_0/(1 + kP_0 t)$.   Therefore
    $P(t) \longrightarrow 0$ as $t \longrightarrow \infty$.

6. The solution of the equation $dP/dt = kP + I$ is

    $$P(t) = (1/k)[(kP_0 + I)e^{kt} - I].$$

    With $P_0 = 1.5$ (population in millions), $k = 0.04$,   and   I
    $= 0.05$ (millions/year), we have

    $$P(t) = 2.75e^{0.04t} - 1.25.$$

The population in the year 2000 will be $P(20) \approx 4.87$ million. With no immigration it would be

$$1.5 \exp [(0.04)(20)] \approx 3.34 \text{ million}.$$

7.   (a) $P(t) = 222e^{0.01t}$, so $P(20) \approx 271.15$ million.

   (b)   Now $P' = 0.01P + 0.5$, and the solution with $P(0) = 222$ is

$$P(t) = 272e^{0.01t} - 50.$$

Hence $P(20) \approx 282.22$ million.

8.   The only difference is that, if $P > M$, then

$$\int \frac{dP}{M - P} = -\ln|M - P| = -\ln(P - M),$$

so integration of the separated equation yields

$$P/(P - M) = Ae^{kMt}.$$

This gives $A = P_0/(P_0 - M)$, so each side of the equation preceding Equation (4) in this section is simply multiplied by -1, and the result in Equation (4) is unchanged.

9.   (a) $x' = 0.8x - 0.004x^2 = 0.004x(200 - x)$, so the maximum amount that will dissolve is $M = 200$ g.

   (b)   With $M = 200$, $P_0 = 50$, and $k = 0.004$, Equation (4) yields the solution

$$x(t) = 10,000/(50 + 150e^{-0.08t}).$$

We find that $x = 100$ when $t = 1.25 \ln 3 \approx 1.37$ sec.

10.   The differential equation for $N(t)$ is

$$N'(t) = kN(15 - N) .$$

When we substitute $N(0) = 5$ (thousands) and $N'(0) = 0.5$ (thousands/day) we find that $k = 0.01$. With N in place of P, this is the logistic equation in (3), so its solution is given by Equation (4):

$$N(t) = \frac{(15)(5)}{5 + 10 \exp[-(0.01)(15)t]} = \frac{15}{1 + 2 \exp(-0.15t)} .$$

From this equation we find that $N = 10$ after $t = (\ln 4)/(0.15) \approx 9.24$ days.

11. Proceeding as in Example 1, we solve the equations

$$25.00k(M - 25.00) = 3/8$$
$$47.54k(M - 47.54) = 1/2$$

for $M = 100$ and $k = 0.0002$. Then Equation (4) gives the population function

$$P(t) = 2500/(25 + 75e^{-0.02t}) .$$

We find that $P = 75$ when $t = 50 \ln 9 \approx 110$, that is, in 2035 A.D.

12. The differential equation for $P(t)$ is

$$P'(t) = 0.001P^2 - \delta P.$$

When we substitute $P(0) = 100$ and $P'(0) = 8$ we find that $\delta = 0.02$, so

$$P'(t) = 0.001P^2 - 0.02P = 0.001P(P - 20) .$$

This is the same as Equation (6) with $k = 0.001$ and $a = 20$, so Equation (7) gives the solution

$$P(t) = (20C\ e^{-0.02t}) / (C\ e^{-0.02t} - 1).$$

$P(0) = 100$ yields $C = 5/4$,   so

$$P(t) = (100\ e^{-0.02t}) / (5\ e^{-0.02t} - 4).$$

From this equation we find that $P = 200$ when $t = 50$
$\ln(9/8) \approx 5.89$ months.

13.   We are given that

$$P' = kP^2 - 0.01P,$$

and the fact that $P = 200$ and $P' = 2$ when $t = 0$
implies that $k = 0.0001$,   so

$$P' = 10^{-4}(P^2 - P).$$

The solution with $P(0) = 200$ is

$$P(t) = 100/(1 - 0.5e^{0.01t}).$$

(a) $P = 1000$ when $t = 100\ \ln(9/5) \approx 58.78$

(b) $P \longrightarrow \infty$ as $t \longrightarrow 100\ \ln 2 \approx 69.31$

14.   The situation here is similar to that in Problem 8, the only
difference being that if $P < a$ then

$$\int \frac{dP}{P - a} = \ln|P - a| = \ln(a - P)\ ,$$

so the second equation preceding Equation (7) in this
section becomes

$$\ln[P/(a - P)] = -\ kat + C_1.$$

16.   If we substitute $P(0) = 10^6$ and $P'(0) = 3 \cdot 10^5$ into the

differential equation

$$P'(t) = \beta_0 e^{-\alpha t} P,$$

we find that $\beta_0 = 0.3$. Hence the solution given in Problem 15 is

$$P(t) = P_0 \exp[(0.3/\alpha)(1 - e^{-\alpha t})].$$

The fact that $P(6) = 2P_0$ now yields the equation

$$f(\alpha) = (0.3)(1 - e^{-6\alpha}) - \alpha \ln 2 = 0$$

for α. We apply Newton's iterative formula

$$\alpha_{n+1} = \alpha_n - f(\alpha_n)/f'(\alpha_n)$$

with $f'(\alpha) = 1.8 \, e^{-6\alpha} - \ln 2$ and initial guess $\alpha_0 = 1$, and find that $\alpha \approx 0.3915$. Therefore the limiting cell population as $t \longrightarrow \infty$ is

$$P_0 \exp(\beta_0/\alpha) = 10^6 \exp(0.3/0.3915) \approx 2.15 \cdot 10^6.$$

Thus the tumor does not grow much further after 6 months.

SECTION 1.9

MOTION WITH VARIABLE ACCELERATION

This section consists of four essentially independent subsections -- resistance proportional to velocity, resistance proportional to velocity squared, inverse-square gravitational acceleration, and rocket propulsion -- that can be studied separately.

The integrations in Problems 2, 4, 6, and 7 are quite elementary with the answers provided in the text, so they are omitted here.

1. Given: $v' = k(250 - v)$, $v(0) = 0$, $v(10) = 100$

    Solution: $v(t) = 250(1 - e^{-kt})$ with $k = (1/10)\ln(5/3)$

    Answer: $v = 200$ when $t \approx 31.5$ sec

3. Given: $v' = -kv$, $v(0) = 40$, $v(10) = 20$

    Solution: $v(t) = 40e^{-kt}$ with $k = (1/10)\ln 2$

    Answer: $x(\infty) = v_0/k = 400/\ln 2 \approx 577$ ft

5. Given: $v' = -kv^2$, $v(0) = 40$, $v(10) = 20$

    Solution: $v(t) = 400/(10 + t)$

    $x(t) = 400 \ln[(10 + t)/10]$

    Answer: $x(60) = 400 \ln 7 \approx 778$ ft

8. (a) The solution of $dv/dt = 18 - 0.12v$ with $v(0) = 0$ is

    $$v(t) = 150(1 - e^{-12t}).$$

    (b) $v(10) \approx 104.82$ ft/sec ≈ 71 mph and $v(\infty) = 150$ ft/sec $\approx$ 102 mph.

9. The solution of the initial value problem

    $$1000v' = 5000 - 100v, \quad v(0) = 0$$

    is

    $$v(t) = 50(1 - e^{-t/10}).$$

    Hence as $t \to \infty$ we see that $v(t)$ approaches $v_{max} = 50$ ft/sec ≈ 34 mph.

10. She opens her parachute at a height of 7089 ft, and reaches the ground in a total time of 5 min 47 sec. For a complete solution see Example 5 in Section 7.6 of Edwards & Penney, *Calculus and Analytic Geometry* (2nd edition, Prentice-Hall 1986).

Section 1.9   43

11. If the paratrooper's terminal velocity was 100 mph = 440/3 ft/sec, then Equation (5) yields $\rho = 12/55$. Then we find by solving Equation (7) numerically with $y_0 = 1200$ and $v_0 = 0$ that $y = 0$ when $t \approx 12.5$ sec. Thus the newspaper account is inaccurate.

12. With $m = 640/32 = 20$ slugs, $W = 640$ lb, $B = (8)(62.5) = 500$ lb, and $F_R = -v$ lb (F_R is upward when $v < 0$), the differential equation is

$$20\, v'(t) = -640 + 500 - v = -140 - v \ .$$

Its solution with $v(0) = 0$ is

$$v(t) = 140(e^{-0.05t} - 1),$$

and integration yields

$$y(t) = 2800(1 - e^{-0.05t}) - 140t.$$

Using these equations we find that $t = 20 \ln(28/13) \approx 15.35$ sec when $v = -5$ ft/sec, and that $y(15.35) \approx -649$ ft. Thus the maximum safe depth is just under 650 ft.

15. Given: $v' = 4 - (v^2/400)$, $v(0) = 0$
Solution: $v(t) = 40(e^{t/5} - 1)/(e^{t/5} + 1)$
Answer: $v(10) \approx 30.46$ ft/sec, $v(\infty) = 40$ ft/sec

18. When we substitute $\rho = 0.0002$, $g = 32$, and $v_0 = 288$ in Equation (19) we get $C \approx 0.6240$. Then we see from Equation (17) that the bolt's maximum height ($v = 0$) occurs when

$$t = C/\sqrt{(\rho g)} \approx 7.8003 \text{ sec,}$$

and Equation (20) gives

$$y_{max} = -(1/\rho)\ln \cos C \approx 1044.14 \text{ ft.}$$

19. From Equation (14) we find that the time required to fall a distance y is

$$t = (\rho g)^{-1/2}\cosh^{-1}(e^{\rho y}).$$

With y = 1044.14 ft (from Problem 18) this gives t ≈ 8.362 sec, and then Equation (13) yields the impact velocity v ≈ 233.72 ft/sec.

20. His terminal speed is $\sqrt{(g/\rho)}$ ≈ 20.66 ft/sec. To find his time of descent, let $\theta = t\sqrt{(\rho g)}$. When y = 10,000 (at the ground), Equation (14) gives

$$\ln \cosh \theta = y\rho = (10000)(0.075) = 750 .$$

Hence θ is certainly large enough that

$$\cosh \theta = (e^{\theta} + e^{-\theta})/2 \approx e^{\theta}/2.$$

Therefore

$$\ln(e^{\theta}/2) \approx 750 ,$$
$$\theta \approx 750 + \ln 2 \approx 750.6931 ,$$
$$t = \theta/\sqrt{(\rho g)} \approx 484.57 \text{ sec,}$$

so the time of descent is about 8 min 5 sec.

22. (a) $0.884 \cdot 10^{-3}$ meters, about 0.88 cm ; (b) $2.91 \cdot 10^{3}$ meters, about 2.91 kilometers. For discussion and details, see the subsection on "black holes" in Section 4.9 of Edwards & Penney, *Calculus and Analytic Geometry* (2nd edition, Prentice-Hall 1986).

23. We get the indicated formula for y_{max} when we set v = 0 in the equation $v^2 = v_0^2 + 2GM(1/y - 1/R)$ preceding Example 3 in this section, solve for y, and finally substitute $GM = gR^2$.

24. (a) We start with

$$v \frac{dv}{dy} = \frac{dv}{dt} = -\frac{k}{y^2}$$

where $k = GM = gR^2$. Integration yields

$$v^2/2 = k(1/y - 1/y_0),$$

since $y = y_0$ when $v = 0$. Solving for $v = dy/dt$ we get

$$\frac{dy}{dt} = -A \left[\frac{y_0 - y}{y} \right]^{1/2}$$

where $A = \sqrt{(2k/y_0)}$. Hence

$$At = \int \left[\frac{y}{y_0 - y} \right]^{1/2} dy$$

$$= \int 2y_0 \cos^2 \theta \; d\theta \qquad\qquad (y = y_0 \cos^2 \theta)$$

$$= y_0(\theta + \sin \theta \cos \theta) \qquad\qquad (t = 0 \quad\text{when}\quad \theta = 0)$$

$$At = y_0[\cos^{-1}\sqrt{(y/y_0)} + (\sqrt{y}/\sqrt{y_0})(\sqrt{(y_0 - y)}/\sqrt{y_0})]$$

The desired formula for t now follows upon division by $A = \sqrt{(2gR^2/y_0)}$.

(b) With $g \approx 79{,}000$ mi/hr^2, $R \approx 3960$ mi, and $y_0 \approx 4960$ mi, the formulas above give $v \approx -11{,}230$ mi/hr and $t \approx 11$ min 33 sec at impact (when $y = R$).

26. $v(t) = -32t + 5280 \ln[25/(25 - t)]$ ft/sec
$y(t) = -16t^2 + 5280 [t - (25 - t)\ln(25/(25 - t))]$ ft
At burnout (when $t = 20$), $y \approx 56{,}711$ ft ≈ 10.74 mi and $v \approx 7858$ ft/sec ≈ 5358 mph.

27. From the rocket propulsion equation $mv' + cm' = -mg$
     (without air resistance) we see that $v' > 0$ (so the rocket
     lifts off) if and only if

$$mg + cm' < 0. \tag{*}$$

The rocket's initial weight is $mg = 50,000$ lb and $m' = -2000/32$ slug/sec, so condition (*) yields $c > 800$ ft/sec.
In this case the rocket lifts off immediately. Its final
weight (after the fuel has all burned) is $mg = 10,000$ lb,
and now condition (*) yields $c > 160$ ft/sec. Thus if 160
$< c \le 800$ then the rocket lifts off sometime during the 20
sec its fuel is burning. If $c < 160$ ft/sec then it never
lifts off.

32. With $m_0 = 878.88$ slugs, $\beta = 8.60$ slug/sec, $c = 6600$
     ft/sec, and $k = 0.1$, Equation (29) becomes

$$\frac{dv}{dt} + \frac{0.1v}{878.88 - 8.60t} = -32.2 + \frac{56763}{878.88 - 8.60t}$$

The integrating factor

$$\rho(t) = (878.88 - 8.60t)^{-0.0116}$$

for this linear first order equation leads to

$$v(t) = 572327 - 32.59t - 529047(878.88 - 8.60t)^{0.0116},$$

$$y(t) = 572327t - 16.30t^2 - 57818717$$
$$+ 60812(878.88 - 8.60t)^{1.0116}.$$

At burnout ($t = 70$ sec) the rocket has velocity $v(70) \approx$
5337 ft/sec $\approx$ 3640 mph and height $y(70) \approx 136,925$ ft $\approx$
25.93 miles.

CHAPTER 1 Review Problems

1.   Linear: $y = x^3(C + \ln x)$

2.   Separable: $y = x/(3 - Cx - x \ln x)$

3.   Homogeneous: $y = x/(C - \ln x)$

4.   Exact: $x^2y^3 + e^x - \cos y = C$

5.   Separable: $y = C \exp[(1 - x)/x^3]$

6.   Separable: $y = x/(1 + Cx + 2x \ln x)$

7.   Linear: $y = x^{-2}(C + \ln x)$

8.   Homogeneous: $y = 3Cx/(C - x^3)$

9.   Bernoulli: $y = (x^2 + C/x)^2$

10.   Separable: $y = \tan(C + x + x^3/3)$

11.   Homogeneous: $y = x/(C - 3 \ln x)$

12.   Exact: $3x^2y^3 + 2xy^4 = C$

13.   Separable: $y = 1/(C + 2x^2 - x^5)$

14.   Homogeneous: $y^2 = x^2/(C + 2 \ln x)$

15.   Linear: $y = (x^3 + C)e^{-3x}$

16.   Substitution: $v = y - x$.
      Solution: $y(1 + Ce^{2x}) = 1 + x + C(1 - x)e^{2x}$

17.   Exact: $e^x + e^y + e^{xy} = C$

18. Homogeneous: $y^2 = Cx^2(x^2 - y^2)$

19. Separable: $y = x^2/(x^5 + Cx^2 + 1)$

20. Linear: $y = 2x^{-3/2} + Cx^{-3}$

21. Linear: $y = [C + \ln(x - 1)]/(x + 1)$

22. Bernoulli: $y = (2x^4 + Cx^2)^3$

23. Exact: $xe^y + y \sin x = C$

24. Separable: $y = x^{1/2}/(6x^2 + Cx^{1/2} + 2)$

25. Linear: $y = (x + 1)^{-2}(x^3 + 3x^2 + 3x + C)$

26. Exact: $6x^{3/2}y^{4/3} - 10x^{6/5}y^{3/2} = C$

27. Bernoulli: $y = x^{-1}(C + \ln x)^{-1/3}$

28. Linear: $y = x^{-1}(C + e^{2x})$

29. Linear: $y = (x^2 + x + C)(2x + 1)^{-1/2}$

30. Substitution: $v = x + y$.
    Solution: $x = 2(x + y)^{1/2} - 2 \ln[1 + (x + y)^{1/2}] + C$

CHAPTER 2

LINEAR EQUATIONS OF HIGHER ORDER

SECTION 2.1

INTRODUCTION

In this section the central ideas of the theory of linear differential equations are introduced and illustrated concretely in the context of *second order* equations. These key concepts include superposition of solutions (Theorem 1), existence and uniqueness of solutions (Theorem 2), linear independence, the Wronskian (Theorem 3), and general solutions (Theorem 4). This discussion of second order equations serves as preparation for the treatment of nth order linear equations in Section 2.2. Although the concepts in this section may seem somewhat abstract to students, the problems set is quite tangible and largely computational.

In each of Problems 1-16 the verification that y_1 and y_2 satisfy the given differential equation is a routine matter. As in Example 2, we then impose the given initial conditions on the general solution $y = c_1y_1 + c_2y_2$. This yields two linear equations that determine the values of the constants c_1 and c_2.

1. $y = 5(e^x - e^{-x})/2$ 2. $y = 2e^{3x} - 3e^{-3x}$
3. $y = 3 \cos 2x + 4 \sin 2x$ 4. $y = 10 \cos 5x - 2 \sin 5x$

5. $y = 2e^x - e^{2x}$ 6. $y = 4e^{2x} + 3e^{-3x}$
7. $y = 6 - 8e^{-x}$ 8. $y = (14 - 2e^{3x})/3$

9. $y = 2e^{-x} + xe^{-x}$ 10. $y = 3e^{5x} - 2xe^{5x}$

11. $y = 5e^x \sin x$

12. If $y(x) = Ae^{-3x}\cos 2x + Be^{-3x}\sin 2x$, then

$$y'(x) = -3e^{-3x}(\ A\cos 2x +\ B\sin 2x)$$
$$+\ e^{-3x}(-2A\sin 2x + 2B\cos 2x).$$

The initial conditions are $y(0) = A = 2$, $y'(0) = -3A + 2B$
$= 0$, so the particular solution is

$$y = e^{-3x}(2\cos 2x + 3\sin 2x).$$

13. $y = 5x - 2x^2$

14. $y = 3x^2 - 16/x^3$

15. $y = 7x - 5x\ln x$

16. $y = 2\cos(\ln x) + 3\sin(\ln x)$

20. Linearly dependent, because

$$f(x)\ =\ \pi\ =\ \pi(\cos^2 x + \sin^2 x)\ =\ \pi\ g(x)$$

21. Linearly independent, because $x^3 = +x^2|x|$ if $x > 0$,
whereas $x^3 = -x^2|x|$ if $x < 0$.

22. Linearly independent, because $1 + x = c(1 + |x|)$ would
require that $c = 1$ with $x = 0$, but $c = 0$ with $x = -1$.
Thus there is no such constant c.

23. Linearly independent, because $f(x) = +g(x)$ if $x > 0$,
whereas $f(x) = -g(x)$ if $x < 0$.

24. Linearly dependent, because $g(x) = 2 f(x)$

25. $f(x) = e^x\sin x$ and $g(x) = e^x\cos x$ are linearly independent, because $f(x) = k\,g(x)$ would imply that $\sin x = k\cos x$, whereas $\sin x$ and $\cos x$ are linearly independent.

26. To see that $f(x)$ and $g(x)$ are linearly independent, assume that $f(x) = c\,g(x)$, and then substitute both $x = 0$ and $x = \pi/2$.

27. Let $L[y] = y'' + py' + qy$. Then $L[y_c] = 0$ and $L[y_p] = f$, so

$$L[y_c + y_p] = L[y_c] + [y_p] = 0 + f = f.$$

28. If $y(x) = 1 + c_1\cos x + c_2\sin x$ then

$$y'(x) = -\,c_1\sin x + c_2\cos x,$$

so the initial conditions $y(0) = y'(0) = -1$ yield $c_1 = -2$, $c_2 = -1$. Hence $y = 1 - 2\cos x - \sin x$.

29. There is no contradiction because if the given differential equation is divided by x^2 to get the form in Equation (8), then the resulting functions $p(x) = -\,4/x$ and $q(x) = 6/x^2$ are not continuous at $x = 0$.

30. (a) $y_1 = x^3$ and $y_2 = |x^3|$ are linearly independent because $x^3 = c|x^3|$ would require that $c = 1$ with $x = 1$, but $c = -1$ with $x = -1$.

(b) The fact that $W(y_1,y_2) = 0$ everywhere does not contradict Theorem 3, because when the given equation is written in the required form

$$y'' - (3/x)y' + (3/x^2)y = 0,$$

the coefficient functions $p(x) = -\,3/x$ and $q(x) = 3/x^2$ are not continuous at $x = 0$.

31. $W(y_1, y_2) = -2x$ vanishes at $x = 0$, whereas if y_1 and y_2 were (linearly independent) solutions of an equation $y'' + py' + qy = 0$ with p and q both continuous on an open interval I containing $x = 0$, then Theorem 3 would imply that $W \neq 0$ on I.

32. (a) $W = y_1y_2' - y_1'y_2$, so

$$AW' = A(y_1'y_2' + y_1y_2'' - y_1''y_2 - y_1'y_2')$$

$$= y_1(Ay_2'') - y_2(Ay_1'')$$

$$= y_1(-By_2' - Cy_2) - y_2(-By_1' - Cy_1)$$

$$= -B(y_1y_2' - y_1'y_2)$$

$$AW' = -BW$$

(b) Just separate the variables.

(c) Because the exponential factor is never zero.

33. $y(x) = c_1e^x + c_2e^{2x}$

34. $y(x) = c_1e^{-5x} + c_2e^{3x}$

35. $y(x) = c_1 + c_2e^{-5x}$

36. $y(x) = c_1 + c_2e^{-3x/2}$

37. $y(x) = c_1e^{-x/2} + c_2e^x$

38. $y(x) = c_1e^{-x/2} + c_2e^{-3x/2}$

39. $y(x) = (c_1 + c_2x)e^{-x/2}$

40. $y(x) = (c_1 + c_2x)e^{2x/3}$

41. $y(x) = c_1e^{-4x/3} + c_2e^{5x/2}$

42. $y(x) = c_1e^{-4x/7} + c_2e^{3x/5}$

GENERAL SOLUTIONS OF LINEAR EQUATIONS

Students should check each of Theorems 1 through 4 in this section to see that, in the case $n = 2$, it reduces to the corresponding theorem in Section 2.1. Similarly, the computational problems for this section largely parallel those for the previous section. By the end of Section 2.2 students should understand that, although we do not prove the existence-uniqueness theorem now, it provides tha basis for everything we do with linear differential equations.

1. $(5/2)(2x) + (-8/3)(3x^2) + (-1)(5x - 8x^2) = 0$

2. $(-4)(5) + (5)(2 - 3x^2) + (1)(10 + 15x^2) = 0$

3. $(1)(0) + (0)(\sin x) + (0)(e^x) = 0$

4. $(1)(17) + (-17/2)(2 \sin^2 x) + (-17/3)(3 \cos^2 x) = 0$

   because $\sin^2 x + \cos^2 x = 1$.

5. $(1)(17) + (-34)(\cos^2 x) + (17)(\cos 2x) = 0$,

   because $2 \cos^2 x = 1 + \cos 2x$.

6. $(-1)(e^x) + (1)(\cosh x) + (1)(\sinh x) = 0$, because

   $\cosh x = (e^x + e^{-x})/2$ and $\sinh x = (e^x - e^{-x})/2$.

7. $$W = \begin{vmatrix} 1 & x & x^2 \\ 0 & 1 & 2x \\ 0 & 0 & 2 \end{vmatrix} = 2 \text{ is nonzero everywhere.}$$

8.
$$W = \begin{vmatrix} e^x & e^{2x} & e^{3x} \\ e^x & 2e^{2x} & 3e^{3x} \\ e^x & 4e^{2x} & 9e^{3x} \end{vmatrix} = 2e^{6x} \quad \text{is nonzero everywhere.}$$

9. $W = e^x(\cos^2 x + \sin^2 x) = e^x \neq 0$

10. $W = x^{-7}e^x(x + 1)(x + 4)$ is nonzero for $x > 0$.

11. $W = x^3 e^{2x}$ is nonzero if $x \neq 0$.

12. $W = x^{-2}[2 \cos^2(\ln x) + 2 \sin^2(\ln x)] = 2x^{-2}$ is nonzero for $x > 0$.

In each of Problems 13-20 we first form the general solution

$$y(x) = c_1 y_1(x) + c_2 y_2(x) + c_3 y_3(x),$$

then calculate $y'(x)$ and $y''(x)$, and finally impose the given initial conditions to determine the values of the coefficients c_1, c_2, c_3.

13. $y = (4e^x - e^{-2x})/3$

14. $y = (3e^x - 6e^{2x} + 3e^{3x})/2$

15. $y = (2 - 2x + x^2)e^x$

16. $y = -12e^x + 13e^{2x} - 10xe^{2x}$

17. $y = 29 - 2 \cos 3x - 3 \sin 3x)/9$

18. $y = e^x(-2 + 2 \cos x + \sin x)$

19. $y = x + 2x^2 + 3x^3$

20. $y = 2x - x^{-2} + x^{-2}\ln x$

In each of Problems 21-24 we first form the general solution

$$y(x) = y_c(x) + y_p(x) = c_1y_1(x) + c_2y_2(x) + y_p(x),$$

then calculate $y'(x)$, and finally impose the given initial conditions to determine the values of the coefficients c_1 and c_2.

21. $y = 2 \cos x - 5 \sin x + 3x$

22. $y = 4e^{2x} - e^{-2x} - 3$

23. $y = e^{-x} + 4e^{3x} - 2$

24. $y = e^x(3 \cos x + 4 \sin x) + x + 1$

25. $L[y] = L[y_1 + y_2] = L[y_1] + L[y_2] = f + g$

26. (a) $y_1 = 2$ and $y_2 = 3x$ (b) $y = 2 + 3x$

27. The equations

$$c_1 + c_2x + c_3x^2 = 0, \quad c_2 + 2c_3x + 0, \quad 2c_3 = 0$$

(the latter two obtained by successive differentiation of the first one) evidently imply that $c_1 = c_2 = c_3 = 0$.

29. If $c_0e^{rx} + c_1xe^{rx} + \cdots + c_nx^ne^{rx} = 0$, then division by e^{rx} yields $c_0 + c_1x + \cdots + c_nx^n = 0$.

30. When the equation $x^2y'' - 2xy' + 2y = 0$ is rewritten in standard form

$$y'' + (-2/x)y' + (2/x^2)y = 0,$$

the coefficient functions $p_1(x) = -2/x$ and $p_2(x) = 2/x^2$ are not continuous at $x = 0$. Thus the hypotheses of Theorem 3 are not satisfied.

31. (b) If y(0) = 1 and y'(0) = 0, then the equation

$$y'' - 2y' - 5y = 0$$

implies that   y''(0) = 2y'(0) + 5y(0) = 5.

32. Let the functions $y_1, y_2, \cdots, y_n$ be chosen as indicated.
     Then evaluation at x = a of the (k - 1)st derivative of
     the equation $c_1 y_1 + c_2 y_2 + \cdots c_n y_n = 0$ yields $c_k = 0$.
     Thus $c_1 = c_2 = \cdots = c_n = 0$, so the functions are linearly
     independent.

33. This follows from the fact that

$$\begin{vmatrix} 1 & 1 & 1 \\ a & b & c \\ a^2 & b^2 & c^2 \end{vmatrix} = (b - a)(c - b)(c - a).$$

34. $W(f_1, f_2, \cdots, f_n) = V \exp(r_i x)$, and neither V nor $\exp(r_i x)$
     vanishes.

SECTION 2.3

HOMOGENEOUS EQUATIONS WITH CONSTANT COEFFICIENTS

This is a purely computational section devoted to the single most
widely applicable type of higher order differential equations --
linear ones with constant coefficients. We include explanatory
comments only when the solution of the characteristic equation is
not routine.

1. $y = c_1 e^{2x} + c_2 e^{-2x}$ 2. $y = c_1 + c_2 e^{3x/2}$

3. $y = c_1 e^{2x} + c_2 e^{-5x}$ 4. $y = c_1 e^{x/2} + c_2 e^{3x}$

5. $y = c_1e^{-3x} + c_2xe^{-3x}$

6. $y = e^{-5x/2}[c_1\cos(x\sqrt{5}/2) + c_2\sin(x\sqrt{5}/2)]$

7. $y = c_1e^{(3x/2)} + c_2xe^{(3x/2)}$

8. $y = e^{3x}(c_1\cos 2x + c_2\sin 2x)$

9. $y = e^{-4x}(c_1\cos 3x + c_2\sin 3x)$

10. $y = c_1 + c_2x + c_3x^2 + c_4e^{-3x/5}$

11. $y = c_1 + c_2x + c_3e^{4x} + c_4xe^{4x}$

12. $y = c_1 + c_2e^x + c_3xe^x + c_4x^2e^x$

13. $y = c_1 + c_2e^{-(2x/3)} + c_3xe^{-(2x/3)}$

14. $y = c_1e^x + c_2e^{-x} + c_3\cos 2x + c_4\sin 2x$

15. $y = c_1e^{2x} + c_2xe^{2x} + c_3e^{-2x} + c_4xe^{-2x}$

16. $y = (c_1 + c_2x)\cos 3x + (c_3 + c_4x)\sin 3x$

17. $y = c_1\cos(x/\sqrt{2}) + c_2\sin(x/\sqrt{2}) + c_3\cos(2x/\sqrt{3}) + c_4\sin(2x/\sqrt{3})$

18. $y = c_1e^{2x} + c_2e^{-2x} + c_3\cos 2x + c_4\sin 2x$

19. $y = c_1e^x + c_2e^{-x} + c_3xe^{-x}$

20. $r^4 + 2r^3 + 3r^2 + 2r + 1 = (r^2 + r + 1)^2$, so

   $y = e^{-x/2}(c_1 + c_2x)\cos(x\sqrt{3}/2) + e^{-x/2}(c_3 + c_4x)\sin(x\sqrt{3}/2)$

21. $y = 5e^x + 2e^{3x}$

22. $y = e^{-x/3}(3\cos x/\sqrt{3} + 5\sqrt{3}\sin x/\sqrt{3})$

23. $y = e^{3x}(3 \cos 4x - 2 \sin 4x)$

24. $y = (-7 + e^{2x} + 8e^{-x/2})/2$

25. $y = (-13 + 6x + 9e^{-2x/3})/4$

26. $y = (24 - 9e^{-5x} - 25xe^{-5x})/5$

27. $y = c_1e^x + c_2e^{-2x} + c_3xe^{-2x}$

28. First we spot the root $r = 2$. Then long division of $2r^3 - r^2 - 5r - 2$ by $r - 2$ yields the quadratic factor

$$2r^2 + 3r + 1 = (2r + 1)(r + 1)$$

with roots $r = -1, -1/2$. Hence the general solution is

$$y = c_1e^{2x} + c_2e^{-x} + c_3e^{-x/2}.$$

29. $y = c_1e^{-3x} + e^{3x/2}[c_2\cos(3x\sqrt{3}/2) + c_3 \sin(3x\sqrt{3}/2)$

30. First we spot the root $r = -1$. Then long division of

$$r^4 - r^3 + r^2 - 3r - 6$$

by $r + 1$ yields the cubic factor $r^3 - 2r^2 + 3r - 6$. Next we spot the root $r = 2$, and another long division yields the factor $r^2 + 3$ with roots $r = \pm i\sqrt{3}$. Hence the general solution is

$$y = c_1e^{-x} + c_2e^{2x} + c_3\cos x\sqrt{3} + c_4\sin x\sqrt{3}.$$

31. $y = c_1e^{0.6527x} + c_2e^{2.8794x} + c_3e^{-0.5321x}$

32. By checking signs, we see that the characteristic equation $r^3 + 3r + 5 = 0$ has a root between -1 and -2. Using Newton's method we find that this root is $r \approx -1.15417150$. Then long division yields the quadratic factor

$$r^2 - 1.15417150\ r + 4.33211185$$

whose approximate roots are $0.57709 \pm 1.99977\ i$. Hence the approximate general solution is

$$y = c_1 e^{-1.15417x} +$$

$$e^{0.57709x}(c_2\cos 1.99977x + c_3\sin 1.99977x).$$

33. $y = c_1 e^{1.2115x} + c_2 e^{-1.9631x}$

$$+ e^{0.3758x}(c_1\cos 1.6739x + c_2\sin 1.6739x)$$

34. Knowing that $y = e^{2x/3}$ is one solution, we divide the characteristic polynomial $3r^3 - 2r^2 + 12r - 8$ by $3r - 2$ and get the quadratic factor $r^2 + 4$. Hence the general solution is

$$y = c_1 e^{2x/3} + c_2\cos 2x + c_3\sin 2x.$$

35. $y = c_1 e^{-x/2} + c_2 e^{-x/3} + c_3\cos 2x + c_4\sin 2x$

36. The fact that $y = e^{-x}\sin x$ is one solution tells us that

$$(r + 1)^2 + 1 = r^2 + 2r + 2$$

is a factor of the characteristic polynomial

$$9r^3 + 11r^2 + 4r - 14.$$

Then long division yields the linear factor $9r - 7$. Hence the general solution is

$$y = c_1 e^{7x/9} + e^{-x}(c_2\cos x + c_3\sin x).$$

38. (a) $x = i, -2i$;   (b) $x = -i, 3i$

39. $y = c_1 e^{-ix} + c_2 e^{3ix}$

40. $y = c_1 e^{3ix} + c_2 e^{-2ix}$

41. $y = c_1 \exp[(1 + i\sqrt{3})x] + c_2 \exp[(-1 - i\sqrt{3})x]$

42. The general solution is

$$y(x) = Ae^x + Be^{\alpha x} + Ce^{\beta x}$$

where $\alpha = (-1 + i\sqrt{3})/2$ and $\beta = (-1 - i\sqrt{3})/2$. Imposition of the given initial conditions yields the equations

$$
\begin{aligned}
A + B + C &= 1 \\
A + \alpha B + \beta C &= 0 \\
A + \alpha^2 B + \beta^2 C &= 0
\end{aligned}
$$

that we solve for $A = B = C = 1/3$.

43. The general solution is

$$y = Ae^{2x} + Be^{-x} + C \cos x + D \sin x.$$

Imposition of the given initial conditions yields the equations

$$
\begin{aligned}
A + B + C &= 0 \\
2A - B \quad + D &= 0 \\
4A + B - C &= 0 \\
8A - B \quad - D &= 30
\end{aligned}
$$

that we solve for $A = 2$, $B = -5$, $C = 3$, and $D = -9$. Thus
$$y = 2e^{2x} - 5e^{-x} + 3 \cos x - 9 \sin x.$$

44. $y_1(x) = \cos x$ if $x \geq 0$, $\cosh x$ if $x \leq 0$.

$y_2(x) = \sin x$ if $x \geq 0$, $\sinh x$ if $x \leq 0$.

MECHANICAL VIBRATIONS

In this section we discuss four types of free motion of a mass on a spring -- undamped, underdamped, critically damped, and overdamped. However, the undamped and underdamped cases -- in which actual oscillations occur -- are emphasized because they are both the most interesting and the most important cases for applications.

1. Frequency: 2 rad/sec = $1/\pi$ Hz; Period: π sec

2. m = 24/32 = 3/4 slug, so the frequency $\omega_0 = [k/m]^{1/2} = [48/(3/4)]^{1/2} = 8$ rad/sec, and the period $T = 2\pi/\omega_0 = \pi/4$ sec.

3. The spring constant is k = 15 N/0.20 m = 75 N/m. The solution of $3x'' + 75x = 0$ with x(0) = 0 and x'(0) = -10 is x(t) = - 2 sin 5t. Thus the amplitude is 2 m; the frequency is 5 rad/sec; and the period is $2\pi/5$ sec.

4. (a) With m = 1/4 slug and k = 3 lb/in = 36 lb/ft we find that $\omega_0 = 12$ rad/sec. The solution of $x'' + 144x = 0$ with x(0) = 1 and x'(0) = -5 is

   x(t) = cos 12t - (5/12)sin 12t

       = (13/12)[(12/13)cos 12t - (5/13)sin 12t]

   x(t) = (13/12)cos(12t - α)

   where $\alpha = 2\pi - \tan^{-1}(5/12) \approx 5.8884$.

   (b) C = 13/12 ≈ 1.0833 ft and $T = 2\pi/12 \approx 0.5236$ sec.

5. The gravitational acceleration at distance R from the center of the earth is $g = GM/R^2$. According to Equation

(6) the (circular) frequency ω of a pendulum is given by $\omega^2 = g/L = GM/R^2L$, so its period is $p = 2\pi/\omega = 2\pi R \sqrt{(L/GM)}$.

6. If the pendulum in the clock executes n cycles per day (86400 sec) at Paris, then its period is $p_1 = 86400/n$ sec. At the equatorial location it takes 24 hr 2 min 40 sec = 86560 sec for the same number of cycles, so its period there is $p_2 = 86560/n$ sec. Now let $R_1 = 3956$ mi be the Earth's "radius" at Paris, and R_2 its "radius" at the equator. Then substitution in the equation $p_1/p_2 = R_1/R_2$ of Problem 5 (with $L_1 = L_2$) yields $R_2 = 3963.33$ mi. Thus this (rather simplistic) calculation gives 7.33 mi as the thickness of the Earth's equatorial bulge.

7. The period equation

$$p = 3960\sqrt{(100.10)} = (3960 + x)\sqrt{100}$$

yields $x \approx 1.9795$ mi $\approx 10,450$ ft for the altitude of the mountain.

8. Let n be the number of cycles of the pendulum required for the clock to register 24 hrs = 86400 sec. Then its period with length $L_1 = 30$ in is $p_1 = 86400/n$ sec. If L_2 denotes its length when its period is $p_2 = 87000/n$ sec, then the equation $p_1/p_2 = \sqrt{L_1}/\sqrt{L_2}$ of Problem 5 (with $R_1 = R_2$) yields $L_2 = 30.42$ in.

10. The $F = ma$ equation $\rho\pi r^2 h x'' = \rho\pi r^2 h g - \pi r^2 x g$ simplifies to

$$x'' + (g/\rho h)x = g.$$

The solution of this equation with $x(0) = x'(0) = 0$ is

$$x(t) = \rho h(1 - \cos \omega_0 t)$$

where $\omega_0 = \sqrt{(g/\rho h)}$. With the given numerical values of ρ, h, and g, the amplitude of oscillation is $\rho h = 100$ cm and the period is $p = 2\pi\sqrt{(\rho h/g)} \approx 2.01$ sec.

11. The fact that the buoy weighs 100 lb means that mg = 100 so m = 100/32 slugs. The weight of water is 62.4 lb/ft^3, so the F = ma equation of Problem 10 is

$$(100/32)x'' = 100 - 62.4\pi r^2 x.$$

It follows that the buoy's circular frequency ω is given by

$$\omega^2 = (32)(62.4\pi)r^2/100.$$

But the fact that the buoy's period is p = 2.5 sec means that $\omega = 2\pi/2.5$. Equating these two results yields $r \approx 0.3173$ ft ≈ 3.8 in.

12. (a) Substitution of $M_r = (r/R)^3 M$ in $F_r = - GM_r m/r^2$ yields
$$F_r = - (GMm/R^3)r.$$

(b) Because $GM/R^3 = g/R$, the equation $mr'' = F_r$ yields the differential equation

$$r'' + (g/R)r = 0.$$

(c) The solution of this equation with r(0) = R and r'(0) = 0 is $r(t) = R \cos \omega_0 t$ where $\omega_0 = \sqrt{(g/R)}$. Hence, with g = 32.2 ft/sec^2 and R = (3960)(5280) ft, we find that the period of the particle's simple harmonic motion is

$$p = 2\pi/\omega_0 = 2\pi\sqrt{(R/g)} \approx 5063.10 \text{ sec} \approx 84.38 \text{ min.}$$

13. Overdamped motion: $x(t) = 4e^{-2t} - 2e^{-4t}$

14. Overdamped motion: $x(t) = 4e^{-3t} - 2e^{-7t}$

15. Critically damped: $x(t) = (5 + 10t)e^{-4t}$

16. Underdamped motion: $x(t) = -2e^{-3t} \sin 4t$
$$= 2e^{-3t}\cos(4t - \pi/2)$$

17. Underdamped motion:
$$x(t) \approx (1/3)\sqrt{(313)}e^{-5t/2}\cos(6t - 0.8254)$$

18. Underdamped motion: $x(t) = e^{-4t}(5 \cos 2t + 12 \sin 2t)$
$$\approx 13e^{-4t} \cos(2t - 1.1760)$$

19. Underdamped motion: $x(t) = e^{-5t}(6 \cos 10t + 8 \sin 10t)$
$$\approx 10e^{-5t} \cos(10t - 0.9273)$$

20. (a) With $m = 12/32 = 3/8$ slug, $c = 3$ lb-sec/ft, and $k = 24$ lb/ft, the differential equation is equivalent to

$$3x'' + 24x' + 192x = 0.$$

The solution with $x(0) = 1$ and $x'(0) = 0$ is

$$x(t) = e^{-4t}[\cos 4t\sqrt{3} + (1/\sqrt{3})\sin 4t\sqrt{3}]$$
$$= (2/\sqrt{3})e^{-4t}[(\sqrt{3}/2)\cos 4t\sqrt{3} + (1/2)\sin 4t\sqrt{3}]$$
$$x(t) = (2/\sqrt{3})e^{-4t}\cos(4t\sqrt{3} - \pi/6).$$

(b) The time-varying amplitude is $2/\sqrt{3} \approx 1.15$ ft; the frequency is $4\sqrt{3} \approx 6.93$ rad/sec; and the phase angle is $\pi/6$.

21. (a) With $m = 100$ slugs we get $\omega = \sqrt{(k/100)}$. But we are given that

$$\omega = (80 \text{ cycles/min})(2\pi)(1 \text{ min}/60 \text{ sec}) = 8\pi/3,$$

and equating the two values yields $k \approx 7018$ lb/ft.

(b) With $\omega_1 = 2\pi(78/60)$ sec^{-1}, Equation (21) in the text yields $C \approx 372.31$ lb/(ft/sec). Hence $p = c/2m \approx 1.8615$. Finally $e^{-pt} = 0.01$ gives $t \approx 2.47$ sec.

28. In the underdamped case we have

$$x(t) = e^{-pt}[A \cos \omega_1 t + B \sin \omega_1 t],$$

$$x'(t) = - pe^{pt}[A \cos \omega_1 t + B \sin \omega_1 t]$$
$$+ e^{-pt}[-A\omega_1 \sin \omega_1 t + B\omega_1 \cos \omega_1 t].$$

The conditions $x(0) = x_0$, $x'(0) = v_0$ yield the equations $A = x_0$ and $- pA + B\omega_1 = v_0$, whence $B = (v_0 + px_0)/\omega_1$.

30. If $x(t) = Ce^{-pt}\cos(\omega_1 t - \alpha)$ then

$$x'(t) = - pCe^{-pt}\cos(\omega_1 t - \alpha) + C\omega_1 e^{-pt}\sin(\omega_1 t - \alpha) = 0$$

yields $\tan(\omega_1 t - \alpha) = - p/\omega_1$.

31. If $x_1 = x(t_1)$ and $x_2 = x(t_2)$ are two successive local maxima, then $\omega_1 t_2 = \omega_1 t_1 + 2\pi$ so

$$x_1 = C \exp(-pt_1) \cos(\omega_1 t_1 - \alpha) \ ,$$
$$x_2 = C \exp(-pt_2) \cos(\omega_1 t_2 - \alpha)$$
$$= C \exp(-pt_2) \cos(\omega_1 t_1 - \alpha).$$

Hence $x_1/x_2 = \exp[-p(t_1 - t_2)]$, and therefore

$$\ln(x_1/x_2) = - p(t_1 - t_2) = 2\pi p/\omega_1.$$

32. With $t_1 = 0.34$ and $t_2 = 1.17$ we first use the equation $\omega_1 t_2 = \omega_1 t_1 + 2\pi$ from Problem 31 to calculate $\omega_1 = 2\pi/(0.83) \approx 7.57$ rad/sec. Next, with $x_1 = 6.73$ and $x_2 =$

1.46, the result of Problem 31 yields

$$p = (1/0.83)\ln(6.73/1.46) \approx 1.84.$$

Then Equation (16) in this section gives

$$c = 2mp = 2(100/32)(1.84) \approx 11.51 \text{ lb-sec/ft,}$$

and finally Equation (21) yields

$$k = (4m^2\omega_1^2 + c^2)/4m \approx 189.68 \text{ lb/ft.}$$

NONHOMOGENEOUS EQUATIONS AND
THE METHOD OF UNDETERMINED COEFFICIENTS

The method of undetermined coefficients is based on "educated guessing". If we can guess correctly the *form* of a particular solution of a nonhomogeneous linear equation with constant coefficients, then we can determine the particular solution explicitly by substitution in the given differential equation. Although it is pointed out at the end of Section 2.5 that this simple approach is not always successful, it does turn out to work well with a surprising number of the nonhomogeneous linear differential equations that arise in elementary scientific applications.

1. $y_p = (1/25)e^{3x}$ 2. $y_p = -(5 + 3x)/2$

3. $y_p = (\cos 3x - 5\sin 3x)/39$

4. $y_p = (-4e^x + 3xe^x)/9$

5.   Substituting $\sin^2 x = (1 - \cos 2x)/2$ on the righthand side leads to

$$y_p = (13 + 3 \cos 2x - 2 \sin 2x)/26.$$

6. $y_p = (4 - 56x + 49x^2)/343$

7. Substituting $\sinh x = (e^x - e^{-x})/2$ on the righthand side
    leads to
    $$y_p = (e^{-x} - e^x)/6 = -(1/3)\sinh x.$$

8. Because $\cosh 2x$ is part of the complementary function, we
    try
    $$y_p = x(A \cosh 2x + B \sinh 2x)$$

    and find that $y_p = (1/4)x \sinh 2x.$

9. $y_c = c_1 e^x + c_2 e^{-3x}$ so we try

    $$y_p = A + x(B + Cx)e^x.$$

    This gives $y_p = -(1/3) + (2x^2 - x)e^x/16.$

10. $y_p = (2x \sin 3x - 3x \cos 3x)/6$

11. $y_p = (3x^2 - 2x)/8$

12. $y_p = Ax + x(B \cos x + C \sin x) = 2x + (1/2)x \sin x$

13. $y_p = e^x(7 \sin x - 4 \cos x)/65$

14. $y_p = x^2(A + Bx)e^x = (-3x^2 e^x + x^3 e^x)/24$

15. $y_p = -17$

16. $y_p = (15 - e^{3x} - 2xe^{3x} + 3x^2 e^{3x})/27$

17. $y_p = (x^2 \sin x - 4x \cos x)/10$

18. $y_p = x(Ae^x) + x(B + Cx)e^{2x}$

$$= -(24xe^x - 19xe^{2x} + 6x^2e^{2x})/144$$

19. $y_p = x^2(A + Bx + Cx^2) = (10x^2 - 4x^3 + x^4)/8$

20. $y_p = -7 + (1/3)xe^x$

21. $y_p = xe^x(A \cos x + B \sin x)$

22. $y_p = Ax^3 + Bx^4 + Cx^5 + Dxe^x$

23. $y_p = (Ax + Bx^2)\cos 2x + (Cx + Dx^2)\sin 2x$

24. $y_p = Ax + Bx^2 + Cxe^{-3x} + Dx^2e^{-3x}$

25. $y_p = (Ax + Bx^2)e^{-x} + (Cx + Dx^2)e^{-2x}$

26. $y_p = e^{3x}[(Ax + Bx^2)\cos 2x + (Cx + Dx^2)\sin 2x]$

27. $y_p = Ax \cos x + Bx \sin x + Cx \cos 2x + Dx \sin 2x$

28. $y_p = (Ax + Bx^2 + Cx^3)\cos 3x + (Dx + Ex^2 + Fx^3)\sin 3x$

29. $y_p = (Ax^3 + Bx^4)e^x + Cxe^{2x} + Dxe^{-2x}$

30. $y_p = (A + Bx + Cx^2)\cos x + (D + Ex + Fx^2)\sin x$

31. $y = \cos 2x + (3/4)\sin 2x + x/2$

32. $y = (15e^{-x} - 16e^{-2x} + e^x)/6$

33. $y = \cos 3x - (2/15)\sin 3x + (1/5)\sin 2x$

34. $y = \cos x - \sin x + (1/2)x \sin x$

35. $y = 1 + x/2 + e^x(4 \cos x - 5 \sin x)/2$

36. $y = (234 + 240x - 12x^2 - 4x^4 - 33e^{2x} - 9e^{-2x})/192$

37. $y = 2 - 2e^x + 2xe^x - (1/2)x^2e^x + (1/6)x^3e^x$

38. $y = (176e^{-x}\cos x + 197e^{-x}\sin x - 6 \cos 3x - 7 \sin 3x)/85$

39. $y = - 3 + 3x - (1/2)x^2 + (1/6)x^3 + 4e^{-x} + xe^{-x}$

40. $y = (5e^x + 5e^{-x} + 10 \cos x - 20)/4$

41. $y_p = 255 - 450x + 30x^2 + 20x^3 + 10x^4 - 4x^5$

42. $y = y_c + y_p$ where y_p is given in Problem 41 and

$$y_c = - 96 \cos x + 288 \sin x + e^{2t} - 160e^{-t}$$

43. (b) $y = c_1\cos 2x + c_2\sin x + (1/4)\cos x - (1/20)\cos 3x$

44. We use the identity

$$\sin x \sin 3x = (\cos 2x - \cos 4x)/2$$

and find that the general solution is

$$y = e^{-x/2}(c_1\cos x\sqrt{3}/2 + c_2\sin x\sqrt{3}/2)$$
$$+ (- 3 \cos 2x + 2 \sin 2x)/26$$
$$+ (-15 \cos 4x + 4 \sin 4x)/482$$

45. We substitute

$$\sin^4 x = (1 - \cos 2x)^2/4$$
$$= (1 - 2 \cos 2x + \cos^2 2x)/4$$
$$= (3 - 4 \cos 2x + \cos 4x)/8$$

on the righthand side. The general solution is

$$y = c_1\cos 3x + c_2\sin 3x$$
$$+ 1/24 - (1/10)\cos 2x - (1/56)\cos 4x.$$

46. The complementary solution is $y_c = c_1\cos x + c_2\sin x$. We

use the identity $\cos^3 x = (\cos 3x + 3 \cos x)/4$ of Problem
43. The particular solution associated with $(1/4)x \cos 3x$
is

$$y_1 = (3 \sin 3x - 4x \cos 3x)/128.$$

The particular solution associated with $(3/4)x \cos x$ is

$$y_2 = (3x \cos x + 3x^2 \sin x)/16.$$

The general solution is $y = y_c + y_1 + y_2$.

47. $y_p = Ax + Bx^2 + Cx^3$ 48. $y_p = Axe^{5x}$

49. $y_p = Axe^{5x} + Bx^2e^{5x}$ 50. $y_p = Ax \cos x + Bx \sin x$

51. $y_p = Ax^2\cos 2x + Bx^2\sin 2x$

52. $y_p = e^x(Ax^2\cos x + Bx^2\sin x)$

SECTION 2.6

REDUCTION OF ORDER AND EULER-CAUCHY EQUATIONS

The key result in this section is the formula

$$y_2 = y_1 \int (y_1)^{-2} \exp\left[- \int p(x) \ dx\right] dx$$

that gives a second solution y_2 in terms of a known solution
y_1 of the *homogeneous* second order linear equation

$$y'' + p(x)y' + q(x)y = 0$$

written in *standard form* (with leading coefficient 1). This formula is applied directly in Problems 1-10, and can also be used to check the answers in Problems 11-17. The subsection on Euler-Cauchy equations is less important and can be omitted. The material on reducible second order equations (with either x or y missing) probably will be more valuable for most students.

1. $p(x) = -4;$ $y_2 = xe^{2x}$

2. Before applying the reduction of order formula, the given differential equation must be written in the standard form

    $$y" + (3/x)y' + (1/x^2)y = 0.$$

    Thus $p(x) = 3/x.$ The answer is $y_2 = (\ln x)/x$.

3. $p(x) = 0;$ $y = x^{1/2}\ln x$

4. $p(x) = (1 - 1/x)$; $y_2 = x - 1$

5. $p(x) = -2(1 + 1/x);$ $y_2 = 2x^2 + 2x + 1$

6. $p(x) = -2x/(x^2 + 1)$; $y_2 = x^2 - 1$

7. $p(x) = -2/x;$ $y_2 = x \sin x$

8. $p(x) = -1/x$; $y_2 = \sin x^2$

9. $p(x) = -(2x + 1)/(x + x^2)$

    $\exp[-\int p(x) \, dx] = x + x^2$

    Then $y_2 = (-1/2)(2x + 1).$ Ignoring the constant factor, we can take $y_2 = 2x + 1.$

10. $\int[-p(x)]dx = \int[3x^2 - 1)/(x^3 - x)]dx = \ln(x^3 - x)$

    $y_2 = x^4 \int x^{-8}(x^3 - x)dx = 1/6x^2 - 1/4$

Because the differential equation is linear we may multiply
by the constant -24 and take $y_2 = 6 - 4/x^2$.

11. $y_2 = xe^{x/2}$ 12. $y_2 = x^4$

13. $y_2 = 1/x^3$ 14. $y_2 = x + 2$

15. $y_2 = xe^x$ 16. $y_2 = x^3 - 2$

17. $y_2 = x^2\cos x$

18. $\displaystyle\int [- p(x)]dx = \int [2x/(1-x^2)]dx = - \ln(1 - x^2)$

$\displaystyle y_2 = x \int [1/x^2(1 - x^2)]dx$

$\displaystyle = (x/2)\int [2/x^2 + 1/(1 - x) + 1/(1 + x)]dx$

$\displaystyle y_2 = - 1 + \frac{x}{2} \ln \frac{1 + x}{1 - x}$

19. $\displaystyle\int [- p(x)]dx = \int (-1/x)dx = - \ln x$

$\displaystyle y_2 = (x^{-1/2}\cos x) \int (x \sec^2 x)(x^{-1})dx$

$\displaystyle = (x^{-1/2}\cos x)(\tan x) = x^{-1/2}\sin x$

20. This problem is for those students who wonder why they spent
so much time in calculus studying techniques of integration.
Noting that $p(x) = - 2x/(1 - x^2)$, we first compute

$$\int [- p(x)]dx = \int [2x/(1 - x^2)] dx = - \ln(1 - x^2).$$

Since $y_1 = 3x^2 - 1$, the reduction of order formula yields

$$y_2 = y_1 \int (3x^2 - 1)^{-2} (1 - x^2)^{-1} \, dx.$$

Observing that only even powers of x are involved, we find the partial fractions decomposition

$$\frac{1}{(1 - x^2)(1 - 3x^2)^2} = \frac{1/4}{1 - x^2} - \frac{3/4}{1 - 3x^2} + \frac{3/2}{(1 - 3x^2)^2} \, .$$

Now in Problem 18 we saw that

$$\int [1/(1 - x^2)] \, dx = (1/2) \ln (1 + x)/(1 - x),$$

and the trigonometric substitution $x\sqrt{3} = \sin \theta$ yields

$$- \frac{3}{4} \int \frac{dx}{1 - 3x^2} + \frac{3}{2} \int \frac{dx}{(1 - 3x^2)^2}$$

$$= (- \sqrt{3}/4) \int \sec \theta \, d\theta + (\sqrt{3}/2) \int \sec^3 \theta \, d\theta$$

$$= (- \sqrt{3}/4) \ln |\sec \theta + \tan \theta|$$
$$\quad + (\sqrt{3}/2)[(1/2) \sec \theta \tan \theta + (1/2) \ln |\sec \theta + \tan \theta|]$$

$$= (\sqrt{3}/4) \sec \theta \tan \theta$$

$$= \frac{3x/4}{1 - 3x^2}$$

because $\sec \theta = 1/(1 - 3x^2)^{1/2}$ and $\tan \theta = x\sqrt{3}/(1 - 3x^2)^{1/2}$. Putting it all together, we finally get

$$y_2 = \frac{1}{8} \left[y_1 \ln \frac{1 + x}{1 - x} - 6x \right] \, .$$

21. $y = c_1 + c_2 \ln x$

22. $y = c_1 x + c_2 x^{-1}$

23. $y = c_1 x^3 + c_2 x^{-4}$

24. $y = c_1 x^2 + c_2 x^2 \ln x$

25. $y = c_1 x^{1/2} + c_2 x^{-3/2}$

26. $y = c_1 x^{4/3} + c_2 x^{-2/3}$

28. $y = x^{-3}[c_1\cos(4\ln x) + c_2\sin(4\ln x)]$

27. $y = x[c_1\cos(\ln x) + c_2\sin(\ln x)]$

29. $y = c_1\cos[\sqrt{(3/2)}\ln x] + c_2\sin[\sqrt{(3/2)}\ln x]$

32. $y = c_1 + c_2\ln x + c_3x^{-3}$

33. $y = c_1 + c_2\ln x + c_3(\ln x)^2$

34. The roots of the characteristic equation
$r^3 - 3r^2 + 2 = 0$ of the transformed equation are $1, 1 \pm \sqrt{3}$.
Hence the general solution of $x^3y^{(3)} - 2xy' + 2y = 0$ is

$$y = c_1x + c_2x^{1+\sqrt{3}} + c_3x^{1-\sqrt{3}}.$$

35. $y = c_1x^{-1} + x^{1/2}\{c_2\cos[\sqrt{(3/4)}\ln x] + c_3\sin[\sqrt{(3/4)}\ln x]\}$

36. The roots of the characteristic equation of the transformed
equation are $2, (-1 \pm i\sqrt{7})/2$. Hence the general solution
of $x^3y^{(3)} + 2x^2y'' - 4y = 0$ is

$$y = c_1x^2 + c_2x^{-1/2}\cos[(\sqrt{7}/2)\ln x] + c_3x^{-1/2}\sin[(\sqrt{7}/2)\ln x].$$

Problems 37-48 involve second order equations in which either the
dependent variable y or the independent variable x is
missing.

37. $y = Ax^2 + B$ 38. $y = (C_1x + C_2)^{1/2}$

39. $y = A\cos 2x + B\sin 2x$ 40. $y = x^2 + C_1\ln x + C_2$

41. $y = A - \ln|x + B|$ 42. $y = \ln x + C_1/x^2 + C_2$

43. $y = (A + Be^x)^{1/2}$

44. $y = \ln|\sec(x + C_1)| - x^2/2 + C_2$

45. $y^3 + 3x + Ay + B = 0$

46. $Ay^2 = 1 + (Ax + B)^2$

47. $y = A \tan(Ax + B)$

48. $y^2 = 1/(Ax + B)$

50. With $p = x'$ and $x'' = pp'$, the equation $x'' + \omega^2 x = 0$ transforms to the separable equation

$$pp' + \omega^2 x = 0$$

whose solution is

$$\frac{dx}{dt} = p = (c_1 - \omega^2 x^2)^{1/2}.$$

Setting $c_1 = k^2 > 0$, it follows that

$$t = \int \frac{dx}{(k^2 - \omega^2 x^2)^{1/2}} \qquad (a = k/\omega)$$

$$= \frac{1}{\omega} \int \frac{dx}{(a^2 - x^2)^{1/2}} = \frac{1}{\omega} \sin^{-1} \frac{x}{a} + C .$$

Hence
$$x = a \sin \omega(t - C)$$
$$= (-a \sin \omega C)\cos \omega t + (a \cos \omega C)\sin \omega t$$
$$x = A \cos \omega t + B \sin \omega t .$$

51. An outline of the steps:

$$Rp' = (1 + p^2)^{3/2}$$

$$\frac{x}{R} = \int \frac{dp}{(1 + p^2)^{3/2}} = \frac{p}{(1 + p^2)^{1/2}} + \frac{a}{R}$$

$$p = \frac{dy}{dx} = \pm \frac{x - a}{[R^2 - (x - a)^2]^{1/2}}$$

$$y = \pm [R^2 - (x - a)^2]^{1/2} + b$$

$$(x - a)^2 + (y - b)^2 = R^2$$

52. An outline of the steps:

$$(x^2 + 1)u'' + 4xu' = 0$$

$$\frac{u''}{u'} = -\frac{4x}{x^2 + 1}$$

$$u' = C/(x^2 + 1)^2$$

$$u = A + B[\tan^{-1}(x) + x/(x^2 + 1)]$$

$$y = (x^2 + 1)u$$

53. An outline of the steps:

$$p(x) = \frac{2x(x - 3)(x + 1)}{x(x - 1)(x + 1)^2} = \frac{4}{x + 1} - \frac{2}{x - 1}$$

$$\exp[-\int p(x)\, dx] = (x - 1)^2/(x + 1)^4$$

Now Formula (8) in the text yields

$$y_2 = \frac{x}{(x + 1)^2} \int \frac{(x - 1)^2}{x^2}\, dx$$

$$= x(x + 1)^{-2} \int (1 - 2/x + 1/x^2)\, dx$$

$$y_2 = (x + 1)^{-2}(x^2 - 1 - 2x \ln x)$$

55. (a) An outline of the steps:

$$u' = -(u^2 + \omega^2)$$

$$(1/\omega)\tan^{-1}(u/\omega) = -x + c_1$$

$$u(x) = \omega \tan \omega(c_1 - x)$$

$$y(x) = c_2\exp[\int \omega \tan \omega(c_1 - x) \, dx]$$

$$= c_2 \cos \omega(c_1 - x)$$

$$y(x) = A \cos \omega x + B \sin \omega x$$

58. With $k = 1$ Equation (37) in the text gives

$$\frac{dy}{dx} = \frac{1}{2}\left[\frac{x}{c} - \frac{c}{x}\right]$$

so

$$y(x) = \frac{1}{2}\left[\frac{x^2}{2c} - c \ln x\right] + A.$$

Hence $y(x) \longrightarrow \infty$ as $x \longrightarrow 0$. Therefore the dog never reaches the y-axis where the cat is.

SECTION 2.7

VARIATION OF PARAMETERS

When it succeeds, the method of undetermined coefficients (Section 2.5) is generally the simplest way of finding a particular solution of a given nonhomogeneous linear equation. However, variation of parameters enjoys the advantage that (in principle) it can be used to find a particular solution of *any* nonhomogeneous linear equation whose complementary function is known. Moreover, its power of expressing a particular solution in terms of an arbitrary nonhomogeneous term $f(x)$ -- as in Problems 31 through 36 -- is of considerable significance, both practical and theoretical.

1. $y_1 = e^{-2x}$,　$y_2 = e^{-x}$,　$W = e^{-3x}$

   $u_1 = -(4/3)e^{3x}$,　$u_2 = 2e^{2x}$,

   $y_p = (2/3)e^x$

2. $y_1 = e^{-2x}$,　$y_2 = e^{4x}$,　$W = 6e^{2x}$

   $u_1 = -x/2$,　$u_2 = -e^{-6x}/12$,

   $y_p = -(6x + 1)e^{-2x}/12$

3. $y_1 = e^{2x}$,　$y_2 = xe^{2x}$,　$W = e^{4x}$

   $u_1 = -x^2$,　$u_2 = 2x$,

   $y_p = x^2 e^{2x}$

4. The complementary function is $y_1 = c_1\cosh 2x + c_2 \sinh 2x$, so the Wronskian is

   $$W = 2 \cosh^2 2x - 2 \sinh^2 2x = 2,$$

   so Formula (19) yields

   $$y_p = -(\cosh 2x)\int (1/2)(\sinh 2x)(\sinh 2x)\ dx$$
   $$+ (\sinh 2x)\int (1/2)(\cosh 2x)(\sinh 2x)\ dx.$$

   Using the identities $2 \sinh^2\theta = \cosh 2\theta - 1$ and

2 sinh θ cosh θ = sinh 2θ, we evaluate the integrals and find that

$$y_p = (4x \cosh 2x - \sinh 4x \cosh 2x + \cosh 4x \sinh 2x)/16$$
$$y_p = (4x \cosh 2x - \sinh 2x)/16$$

5. $y_p = -(1/4)(\cos 2x \cos x - \sin 2x \sin x)$
$$+ (1/20)(\cos 5x \cos 2x + \sin 5x \sin 2x)$$

$$= -(1/5)\cos 3x \quad (!)$$

6. $y_1 = \cos 3x, \quad y_2 = \sin 3x, \quad W = 3$

$y_p = (\sin 3x - 6x \cos 3x)/36$

7. $y_1 = \cos 3x, \quad y_2 = \sin 3x, \quad W = 3$

$u_1' = -(2/3) \tan 3x, \quad u_2' = 2/3$

$y_p = (2/9)[3x \sin 3x + (\cos 3x)\ln|\cos 3x|]$

8. $y_1 = \cos x, \quad y_2 = \sin x, \quad W = 1$

$u_1' = - \csc x, \quad u_2' = \cos x \csc^2 x$

$y_p = - 1 - (\cos x)\ln|\csc x - \cot x|$

9. $y_1 = \cos 2x, \quad y_2 = \sin 2x, \quad W = 2$

$u_1' = -(1/2)\sin^2 x \sin 2x = -(1/4)(1 - \cos 2x)\sin 2x$

$u_2' = (1/2)\sin^2 x \cos 2x = (1/4)(1 - \cos 2x)\cos 2x$

$y_p = (1 - x \sin 2x)/8$

10. $y_1 = e^{-2x}, \quad y_2 = e^{2x}, \quad W = 4$

$$u_1 = -(3x - 1)e^{3x}/36, \qquad u_2 = -(x + 1)e^{-x}/4$$

$$y_p = -e^x(3x + 2)/9$$

11. $y_1 = e^x, \qquad y_2 = xe^x, \qquad W = e^{2x}$

$$u_1 = -\ln x, \qquad u_2 = -1/x$$

$$y_p = -e^x(1 + \ln x)$$

12.   Here it is important to remember that the differential
      equations in this section are written in standard form with
      leading coefficient 1. We therefore rewrite the given
      Euler-Cauchy equation with complementary function $y_c = c_1 x^2$
      $+ c_2 x^3$ as

$$y'' - (4/x)y' + (6/x^2)y = x.$$

Thus $f(x) = x$, and $W = x^4$, so Formula (19) yields

$$y_p = -x^2 \int x^3 \cdot x \cdot x^{-4} dx + x^3 \int x^2 \cdot x \cdot x^{-4} dx$$

$$= -x^2 \int dx + x^3 \int (1/x)dx = x^3(\ln x - 1).$$

13. $y_1 = x^2, \qquad y_2 = x^2 \ln x, \qquad W = x^3, \qquad f(x) = x^2$

$$u_1' = -x \ln x, \qquad u_2' = x$$

$$y_p = x^4/4$$

14. $y_c = c_1 x^{1/2} + c_2 x^{3/2}$; $f(x) = 2x^{-2/3}$; $y_p = -72x^{4/3}/5$

15. $y_1 = \cos(\ln x), \qquad y_2 = \sin(\ln x), \qquad W = 1/x,$

$$f(x) = (\ln x)/x^2$$

$$u_1 = (\ln x)\cos(\ln x) - \sin(\ln x)$$

$$u_2 = (\ln x)\sin(\ln x) + \cos(\ln x)$$

$$y_p = \ln x \quad (!)$$

16. $y_1 = 1, \quad y_2 = e^{2x}, \quad y_3 = e^{-x}, \quad W = 6e^x$

$$u_1' = -x^2/2, \quad u_2' = x^2e^{-2x}/6, \quad u_3' = x^2e^x/3$$

$$y_p = -(9x - 3x^2 + 2x^3)/12$$

17. $y_1 = e^{-x}, \quad y_2 = xe^{-x}, \quad y_3 = x^2e^{-x}$

$$y_p = (1/6)x^3e^{-x}$$

18. $y_1 = 1, \quad y_2 = \cos 2x, \quad y_3 = \sin 2x, \quad W = 8$

$$u_1' = (\cot 2x)/4, \quad u_2' = -(\cos^2 2x)(4 \sin 2x),$$

$$u_3' = -(\cos 2x)/4$$

$$y_p = [-1 + \ln|\sin 2x| - (\cos 2x)\ln|\csc 2x - \cot 2x|]/8$$

19. $y_1 = e^{-x}, \quad y_2 = e^x, \quad W = 2$

$$y_p = \frac{1}{2} e^x(1 - x^{-1}) - \frac{1}{2} e^{-x} \int_1^x t^{-2} e^{2t} \, dt$$

20. $y_1 = \cos x, \quad y_2 = \sin x, \quad W = 1$

$$u_1' = -x^{1/2}\sin x, \quad u_2' = x^{1/2}\cos x$$

$$y_p(x) = \int_0^x t^{1/2}\sin(x - t) \, dt$$

21. $y_p = \frac{1}{2} \int_0^x \exp(-t^2) \sin 2(x - t)\ dt$

22. $y_1 = e^{-x}$, $\quad y_2 = \cos x$, $\quad y_3 = \sin x$, $\quad W = 2e^{-x}$

$u_1' = e^x/2x$, $\quad u_2' = -(\cos x + \sin x)/2x$,

$u_3' = (\cos x - \sin x)/2x$

$y_p(x) = \int_1^x (1/2t)[\exp(t - x) - \cos(t - x) - \sin(t - x)]\ dt$

23. $y_1 = 1$, $\quad y_2 = x$, $\quad y_3 = e^x$, $\quad W = e^x$

$u_1' = (x - 1)\ln x$, $\quad u_2' = -\ln x$, $\quad u_3' = e^{-x}\ln x$

$y_p = \dfrac{3x^2}{4} + \dfrac{1}{2}(x^2 - 4x)\ln x + e^x \int_1^x e^{-t}\ln t\ dt$

24. The complementary function is

$$y_c = c_1\cos x + c_2\sin x + c_3\cosh x + c_4\sinh x\ .$$

When we set up the Wronskian determinant, then subtract the first row from the third row and the second row from the fourth row, we find that $W = 4$. Similar determinant computations yield

$u_1' = (1/2)\sin x \tanh x$, $\quad u_2' = -(1/2)\cos x \tanh x$,
$u_3' = -(1/2)\sinh x \tanh x$, $\quad u_4' = (1/2)\cosh x \tanh x$.

These derivatives and the trigonometric and hyperbolic addition formulas lead to the particular solution

$$y_p(x) = \frac{1}{2} \int_0^x (\tanh t)[\sin(t - x) - \sinh(t - x)] \, dt.$$

25. $y_1 = x, \quad y_2 = x^{-1}, \quad y_3 = x^{-2}, \quad W = -6x^{-5}, \quad f(x) = x$

$u_1' = x^3/18, \quad u_2' = -x^5/10, \quad u_3' = x^6/18$

$y_p = x^4/90$

26. The Wronskian of the homogeneous solutions

$$y_1 = x^{1/2}, \quad y_2 = x^{1/3}, \quad y_3 = x^{1/4}$$

is $W = -288x^{-23/12}$. We find that

$$u_1' = 24x^{3/2}, \quad u_2' = -72x^{5/3}, \quad u_3' = 48x^{7/4}.$$

The resulting particular solution is $y_p = 3x^3/55$. Of course we might well have noted by inspection that there should be a particular solution of the form $y_p = Ax^3$. Having done so, A could be found by the method of undetermined coefficients.

27. $y_1 = x, \quad y_2 = 1 + x^2, \quad W = x^2 - 1, \quad f(x) = 1$

$u_1' = (1 + x^2)/(1 - x^2), \quad u_2' = x/(x^2 - 1)$

$y_p = -x^2 + x \ln|(1 + x)/(1 - x)| + (1/2)(1 + x^2)\ln|1 - x^2|$

28. The Wronskian of $y_1 = x^{-1/2}\cos x$ and $y_2 = x^{-1/2}\sin x$ is $W = 1/x$. We then find that

$$u_1 = (\cos^2 x)/2 \quad \text{and} \quad u_2 = (x + \sin x \cos x)/2.$$

The resulting particular solution is

$$y_p = (x^{1/2}\sin x + x^{-1/2}\cos x)/2.$$

Noting that $(x^{-1/2}\cos x)/2$ satisfies the associated homogeneous equation, we conclude that $y = (x^{1/2}\sin x)/2$ is a particular solution of the given nonhomogeneous equation.

29. $y_1 = e^{-x}$, $\quad y_2 = e^x$, $\quad W = 2$

$u_1' = -x^{-1}e^x/2$, $\quad u_2' = x^{-1}e^{-x}/2$

When we impose the conditions $y(1) = y'(1) = 0$ on the general solution

$$y = Ae^{-x} + Be^x + u_1y_1 + u_2y_2,$$

we find that $A = 0$ and $B = e^{-1}$. Hence the desired particular solution is

$$y = e^{x-1} + \frac{1}{2}e^x\int_1^x t^{-1}e^{-t}dt - \frac{1}{2}e^{-x}\int_1^x t^{-1}e^t dt.$$

30. The Wronskian of $y_1 = e^x$ and $y_2 = e^{2x}$ is $W = e^{3x}$. We find that

$$u_1' = -e^{-x}\sin x^2 \quad \text{and} \quad u_2' = e^{-2x}\sin x^2,$$

so the general solution of the given equation is

$$y = c_1e^x + c_2e^{2x} - e^x\int_0^x e^{-t}\sin t^2 dt + e^{2x}\int_0^x e^{-2t}\sin t^2 dt.$$

When we impose the initial conditions $y(0) = 2$ and $y'(0) = 3$, we find that $c_1 = c_2 = 1$.

31. This is the special case $\alpha = 0$, $\beta = 1$ of Problem 37 below.

32. This is the special case $r_1 = 1$, $r_2 = -1$ of Problem 35 below.

33. If we write Equation (19) with the upper and lower limits given in (20), and use t for the dummy variable of integration, we get

$$y(x) = \int_{x_0}^{x} \frac{y_1(t)y_2(x) - y_2(t)y_1(x)}{W(t)} f(t) \ dt.$$

It is immediate that $y_p(x_0) = 0$, and then the fundamental theorem of calculus yields $y_p'(x_0) = 0$.

34. Formula (22) for $G(x,t)$ now follows immediately upon inspecting the integral formula given above in Problem 33.

35. If $y_1 = \exp(r_1 t)$ and $y_2 = \exp(r_2 t)$, then $W = (r_2 - r_1)\exp[(r_1 + r_2)t]$.

36. If $y_1 = \exp(r_1 x)$ and $y_2 = x \exp(r_1 x)$, then $W = \exp(2r_1 x)$.

37. If $y_1 = e^{\alpha x}\cos \beta x$ and $y_2 = e^{\alpha x}\sin \beta x$, then it follows that $W = \beta e^{2\alpha x}$. Hence

$$G(x,t) = \frac{1}{\beta e^{2\alpha t}} \begin{vmatrix} e^{\alpha t}\cos \beta t & e^{\alpha x}\cos \beta x \\ e^{\alpha t}\sin \beta t & e^{\alpha x}\sin \beta x \end{vmatrix}$$

$$= \beta^{-1}e^{-2\alpha t}[\sin \beta x \cos \beta t - \cos \beta x \sin \beta t]e^{\alpha t}e^{\alpha x}$$

$$= \beta^{-1}e^{\alpha(x-t)}\sin \beta(x - t)$$

as desired.

FORCED OSCILLATIONS AND RESONANCE

1. Trial of $x = A \cos 2t$ yields the particular solution $x_p = 2 \cos 2t$. Hence the general solution is

$$x(t) = c_1 \cos 3t + c_2 \sin 3t + 2 \cos 2t.$$

The initial conditions imply that $c_1 = -2$ and $c_2 = 0$, so $x = 2 \cos 2t - 2 \cos 3t$.

2. First we apply the method of undetermined coefficients to find the particular solution $x_p = - \sin 3t$. Then we impose the initial conditions $x(0) = x'(0) = 0$ on the general solution

$$x(t) = c_1 \cos 2t + c_2 \sin 2t - \sin 3t,$$

and find that $x = (3/2) \sin 2t - \sin 3t$.

3. First we apply the method of undetermined coefficients to find the particular solution

$$x_p = (3/15) \cos 5t + (4/5) \sin 5t$$
$$= (1/3) \cos(5t - \beta)$$

where $\beta = \tan^{-1}(4/3) \approx 0.9273$. Hence the general solution is

$$x(t) = c_1 \cos 10t + c_2 \sin 10t$$
$$+ (3/15) \cos 5t + (4/15) \sin 5t.$$

The initial conditions $x(0) = 25$, $x'(0) = 0$ now yield $c_1 = 372/15$ and $c_2 = -2/15$, so the part of the solution with frequency $\omega = 10$ is

$$x_c = (1/15)(372 \cos 10t - 2 \sin 10t)$$

$$= (1/15)\sqrt{(138,388)} \cos(10t - \alpha)$$

where $\alpha = 2\pi - \tan^{-1}(1/186) \approx 6.2778$.

4. $x = [(-10/9)\cos 5t + 2 \sin 5t] + (10/9)\cos 4t$
    $= (2/9)\sqrt{106} \cos(5t - \alpha) + (10/9)\cos 4t$

where $\alpha = \pi - \tan^{-1}(9/5) \approx 2.07789$

5. $x(t) = (x_0 - C)\cos \omega_0 t + C \cos \omega t$ where $C = F_0/(k - m\omega^2)$

6. $x(t) = [(2mv_0\omega_0 + F_0)/2m\omega_0^2]\sin \omega_0 t - (F_0/2m\omega_0)t \cos \omega_0 t$

7. When we substitute $x = A \cos 3t + B \sin 3t$ in the given differential equation we get the equations

$$- 5A + 12B = 10, \quad - 12A - 5B = 0$$

with solution $A = - 50/169$, $B = 120/169$. Hence

$$x_{sp}(t) = (10/169)(- 5 \cos 3t + 12 \sin 3t)$$
$$= (10/13)\cos(3t - \alpha)$$

where $\alpha = \pi - \tan^{-1}(12/5) \approx 1.9656$.

8. $x_{sp} = (4/25)\cos(4t - \alpha); \quad \alpha = 2\pi - \tan^{-1}(3/4) \approx 5.6397$

9. $x_{sp} = (3/\sqrt{40,001})\cos(10t - \alpha)$
    where $\alpha = \pi + \tan^{-1}(199/20) \approx 4.6122$

10. $x_{sp} \approx 0.09849 \cos(10t - \alpha)$
    where $\alpha = \pi + \tan^{-1}(171/478) \approx 3.4851$

11. $x_{sp} = (1/4)\sqrt{(10)} \cos(3t - \alpha)$
    where $\alpha = \pi - \tan^{-1}3 \approx 1.8925$

$$x_{tr} = (1/4)\sqrt{(50)}\,e^{-2t}\cos(t - \beta)$$

where $\beta = 2\pi - \tan^{-1}7 \approx 4.8543$

12. $x = (75/522)e^{-3t}(2 \cos 2t + 5 \sin 2t)$

$+ (15/261)(- 5 \cos 5t - 2 \sin 5t)$

$\approx 0.77373\, e^{-3t}\cos(2t - 1.19029)$

$+ 0.30949 \cos(5t - 3.52210)$

13. $x_{sp} = (3/\sqrt{9236})\cos(10t - \alpha),\quad \alpha = \pi - \tan^{-1}(10/47) \approx 2.9288$

$x_{tr} \approx 10.9761\, e^{-t}\cos(t\sqrt{5} - 0.4181)$

14. $x = (1/40)e^{-4t}(199 \cos 3t + 258 \sin 3t)$

$+ (1/40)(\cos t + 22 \sin t)$

$\approx 8.14574\, e^{-4t}\cos(3t - 0.91379)$

$+ 0.55057 \cos(t - 1.52537)$

15. $m = 100/32$ slugs and $k = 1200$ lb/ft

so $\omega_0 = \sqrt{(k/m)} = \sqrt{(384)}$ rad/sec ≈ 3.12 Hz.

16. Let the machine have mass m. Then the force $F = mg$ (the machine's weight) causes a displacement of 1/48 ft, so the spring constant is $k = 48mg$ lb/ft. Hence the resonance frequency is

$$\omega = \sqrt{(k/m)} = \sqrt{(48g)}$$
$$= \sqrt{(48 \cdot 32)} = 16\sqrt{6} \text{ rad/sec} \approx 374 \text{ rpm.}$$

17. If θ is the angular displacement from the vertical, then the (essentially horizontal) displacement of the mass is x $= L\theta$, so twice its total energy (KE + PE) is

$$m(x')^2 + kx^2 + 2\,mgh$$

$$= mL^2(\theta')^2 + kL^2\theta^2 + 2\ mgL(1 - \cos\ \theta) = C.$$

Differentiation, substitution of $\theta \approx \sin\ \theta$, and simplification yields

$$\theta'' + (k/m + g/L)\theta = 0$$

so

$$\omega_0 = (k/m + g/L)^{1/2}.$$

18. Let x denote the displacement of the mass from its equilibrium position, $v = x'$ its velocity, and $\omega = v/a$ the angular velocity of the pulley. Then conservation of energy yields

$$mv^2/2 + I\omega^2/2 + kx^2/2 - mgx = C.$$

When we differentiate both sides with respect to t and simplify the result, we get the differential equation

$$(m + I/a^2)x'' + kx = mg\ .$$

Hence $\omega = [k/(m + I/a^2)]^{1/2}.$

19. (a) In ft-lb-sec units we have $m = 1000$ and $k = 10000$, so $\omega_0 = \sqrt{10}$ rad/sec ≈ 0.50 Hz.

(b) We are given that $\omega = 2\pi/2.25 \approx 2.79$ rad/sec, and the equation $mx'' + kx = F(t)$ simplifies to

$$x'' + 10x = (1/4)\omega^2\sin\ \omega t.$$

When we substitute $x(t) = A \sin\ \omega t$ we find that the amplitude is

$$A = \omega^2/4(10 - \omega^2) \approx 0.8854 \text{ ft} \approx 10.63 \text{ in.}$$

20. By the identity of Problem 43 in Section 2.5, the differential equation is

$$mx'' + kx = F_0(3 \cos \omega t + \cos 3\omega t)/4.$$

Hence resonance occurs when either ω or 3ω equals $\omega_0 = \sqrt{(k/m)}$, that is, when either $\omega = \omega_0$ or $\omega = \omega_0/3$.

22. Let $G_0 = (E_0^2 + F_0^2)^{1/2}$. Then

$$
\begin{aligned}
x_{sp} &= (\rho E_0/k)\cos(\omega t - \alpha) + (\rho F_0/k)\sin(\omega t - \alpha) \\
&= (\rho G_0/k)[(E_0/G_0)\cos(\omega t - \alpha) + (F_0/G_0)\sin(\omega t - \alpha)] \\
&= (\rho G_0/k)[\cos \beta \cos(\omega t - \alpha) + \sin \beta \sin(\omega t - \alpha)] \\
x_{sp} &= (\rho G_0/k)\cos(\omega t - \alpha - \beta)
\end{aligned}
$$

where $\tan \beta = F_0/E_0$. Equation (24) now results when we substitute the value of ρ given in Equation (19).

24. Equation (25) gives the amplitude C in terms of $m = 50$, $c = 200$, $k = 4800$ and the circular frequency ω. Substitution of these numerical values yields

$$\frac{c^2}{a^2} = \frac{16\omega^2 + 9216}{(96 - \omega^2)^2 + 16\omega^2}.$$

To find the maximum amplitude C, we differentiate with respect to ω and set the result equal to zero. After considerable simplification we wind up with the equation

$$\omega^4 + 1152\ \omega^2 - 110592 = 0.$$

The only positive real root is $\omega \approx 9.44$ rad/sec. Since $\omega = 2\pi v/30$, this frequency corresponds to velocity

$$v = (30 \cdot 9.44)/2\pi \approx 45.07 \text{ ft/sec} \approx 30.73 \text{ mph}.$$

The initial equation above now yields (with $a = 2$ inches) a maximal amplitude of $c \approx 5.38$ inches at this velocity.

25. The given differential equation corresponds to Equation (17) with $F_0 = mA\omega^2$. It therefore follows from Equations (19) and (20) that the amplitude of the steady periodic vibrations at frequency ω is

$$\frac{F_0}{[(k - m\omega^2)^2 + (c\omega)^2]^{1/2}} = \frac{mA}{k} \cdot \frac{k\omega^2}{[(k - m\omega^2)^2 + (c\omega)^2]^{1/2}} \cdot$$

SECTION 2.9

ELECTRICAL CIRCUITS

1. $5I' + 25I = 0,$ $I(0) = 4;$ $I(t) = 4e^{-5t}$

2. $5I' + 25I = 100,$ $I(0) = 0;$ $I(t) = 4(1 - e^{-5t})$

3. $5I' + 25I = 100 \cos 60t,$ $I(0) = 0$

Substitution of the trial solution

$\qquad I_p = A \cos 60t + B \sin 60t$
yields
$\qquad I_p = 4(\cos 60t + 12 \sin 60t)/145.$

The complementary function is $I_c = ce^{-5t}$; the solution with $I(0) = 0$ is

$\qquad I(t) = 4(\cos 60t + 12 \sin 60t - e^{-5t})/145.$

4. $I(t) = 5(e^{-10t} - e^{-20t})$; we find that $I'(t) = 0$ when $t = (\ln 2)/10,$ and then $I_{max} = 5/4.$

5. $2I' + 20I = 100e^{-10t}\cos 60t$, $I(0) = 0$

    $I(t) = (5/6)e^{-10t}\sin 60t$

6. $I_{sp} = (1/37)(-21 \cos 60t + 22 \sin 60t)$

    $= (5/\sqrt{37})\cos(60t - \alpha)$ where $\alpha = \pi - \tan^{-1}(22/21)$

7. $RQ' + (1/C)Q = E_0$, $Q(0) = 0$

    $Q(t) = E_0C(1 - e^{-t/RC})$

    $I(t) = Q'(t) = (E_0/R)e^{-t/RC}$

8. (a) $Q(t) = 10te^{-5t}$ and $I(t) = 10(1 - 5t)e^{-5t}$.

    (b) $I(t) = 0$ when $t = 1/5$, so $Q_{max} = 2e^{-1}$.

9. $Q(t) = (\cos 120t + 6 \sin 120t - e^{-20t})/1480$

    $I(t) = (36 \cos 120t - 6 \sin 120t + e^{-20t})/74$

    Steady state amplitude $= 3/\sqrt{37}$

11. $I_{sp} = (1/37)(10 \cos 2t + 60 \sin 2t)$

    $= (10/\sqrt{37})\sin(2t - \delta)$ where $\delta = 2\pi - \tan^{-1}(1/6)$

12. $I_{sp} = (2/17)(\cos 10t + 4 \sin 10t)$

    $= (2/\sqrt{17})\sin(10t - \delta)$ where $\delta = 2\pi - \tan^{-1}(1/4)$

13. $I_{sp} = (20/13)(2 \cos 5t + 3 \sin 5t)$

    $= (20/\sqrt{13})\sin(5t - \delta)$ where $\delta = 2\pi - \tan^{-1}(2/3)$

14. Substitution of $I_{sp} = A \cos 100t + B \sin 100t$ in

    $I'' + 10I' + 40I = -6000 \sin 100t + 8000 \cos 100t$

Section 2.9

leads to the equations

$$-9.96A + \quad B = 8$$
$$- \quad A - 9.96B = -6,$$

whence we find that

$$I_{sp} \approx 0.67624 \sin 100t - 0.73532 \cos 100t$$
$$\approx 0.99899 \sin(100t - 0.82723)$$

15. $I_{sp} = I_0 \sin(60\pi t - \delta)$ where
$I_0 = 33\pi[(1000 - 36\pi^2)^2 + (30\pi)^2]^{-1/2} \approx 0.1591$
$\delta = 2\pi - \tan^{-1}[(1000 - 36\pi^2)/30\pi] \approx 4.8576$

16. $I \approx 0.54168 \cos 377t + 1.51877 \sin 377t$
$\approx 1.61247 \sin(377t + 0.34259)$

17. With $I(0) = 0$ and $Q(0) = 5$, Equation (16) in the text
gives $I'(0) = -75$. The solution of $2I'' + 16I' + 50I = 0$
with these initial conditions is $I(t) = -25e^{-4t}\sin 3t$.

18. Our differential equation to solve is

$$2I'' + 60I' + 400I = -100e^{-t}.$$

We find the particular solution $I_p = (-50/171)e^{-t}$ by
undetermined coefficients; the general solution is

$$I(t) = c_1 e^{-10t} + c_2 e^{-20t} - (50/171)e^{-t}.$$

The initial conditions are $I(0) = 0$ and $I'(0) = 50$, the
latter found by substituting $L = 2$, $R = 60$, $1/C = 400$,
$I(0) = Q(0) = 0$, and $E(0) = 100$ into Equation (16).
Imposition of these initial values on the general solution
above yields

$$I(t) = (50/171)(19e^{-10t} - 18e^{-20t} - e^{-t}).$$

19. $2I'' + 60I' + 400I = -1000e^{-10t}$

$I(0) = 0, \quad I'(0) = -150$

$I(t) = 10e^{-20t} - 10e^{-10t} - 50te^{-10t}$

20. The differential equation is $10I'' + 30I' + 50I = 100 \cos 2t$, and the initial conditions are $I(0) = I'(0) = 0$. First we find separately the transient and steady periodic solutions

$$I_{tr} = e^{-3t/2}[c_1\cos (t/2)\sqrt{11} + c_2\sin (t/2)\sqrt{11}],$$

$$I_{sp} \approx 0.27027 \cos 2t + 1.62162 \sin 2t.$$

Then we impose the initial conditions and get

$$I(t) \approx 2.21676\, e^{-3t/2}\sin[(t/2)\sqrt{11} - 3.01937]$$
$$+ 1.64399 \sin(2t + 0.16515).$$

21. $10I'' + 20I' + 100I = -1000 \sin 5t$

$I(0) = 0, \quad I'(0) = -10$

$I(t) = (20/13)(2 \cos 5t + 3 \sin 5t)$
$\qquad\quad - (10/39)e^{-t}(12 \cos 3t + 47 \sin 3t)$

22. $I(t) \approx 0.15961\, e^{-25t}\sin[25t\sqrt{159} + 4.54718]$
$\qquad\quad + 0.15912 \sin(60\pi t + 1.42563)$

23. Critical frequency: $\omega_0 = 1/\sqrt{(LC)}$

24. We need only observe that the roots

$$r = [-R \pm (R^2 - 4L/C)^{1/2}]/2L$$

necessarily have *negative* real parts.

ENDPOINT PROBLEMS AND EIGENVALUES

The material on eigenvalues and endpoint problems in Section 2.10 is optional in a first course, and will not be needed until we discuss boundary value problems in the last three sections of Chapter 8 and in Chapter 9. However, after the concentration thusfar on initial value problems, the inclusion of this section can give students a view of a new class of problems that has diverse and important applications (as illustrated by the subsection on the whirling string). If Section 2.10 is not covered at this point in the course, then it can be inserted just prior to Section 8.5.

1. If $\lambda = 0$ then $y'' = 0$ implies that $y(x) = A + Bx$. The endpoint conditions $y'(0) = 0$ and $y(1) = 0$ yield $B = 0$ and $A = 0$, respectively. Hence $\lambda = 0$ is *not* an eigenvalue

   If $\lambda = \alpha^2 > 0$, then the general solution of $y'' + \alpha^2 y = 0$ is
   $$y(x) = A \cos \alpha x + B \sin \alpha x,$$
   so
   $$y'(x) = - A\alpha \sin \alpha x + B\alpha \cos \alpha x.$$

   Then $y'(0) = 0$ yields $B = 0$, so $y(x) = A \cos \alpha x$. Next $y(1) = 0$ implies that $\cos \alpha = 0$, so α is an odd multiple of $\pi/2$. Hence the positive eigenvalues are $\{(2n - 1)^2\pi^2/4\}$ with associated eigenfunctions $\{\cos(2n - 1)\pi x/2\}$ for $n = 1,2,3,\cdots$.

2. If $\lambda = 0$ then $y'' = 0$ implies that $y(x) = A + Bx$. The endpoint conditions $y'(0) = y'(\pi) = 0$ imply only that $B = 0$, so $\lambda_0 = 0$ is an eigenvalue with associated eigenfunction $y_0(x) = 1$.

If $\lambda = \alpha^2 > 0$, then the general solution of $y'' + \alpha^2 y =$ is

$$y(x) = A \cos \alpha x + B \sin \alpha x.$$

Then

$$y'(x) = -A\alpha \sin \alpha x + B\alpha \cos \alpha x,$$

so $y'(0) = 0$ implies that $B = 0$. Next, $y'(\pi) = 0$ implies that $\alpha\pi$ is an integral multiple of π. Hence the positive eigenvalues are $\{n^2\}$ with associated eigenfunctions $\{\cos nx\}$, $n = 1, 2, 3, \cdots$.

3. Much as in Problem 1 we see that $\lambda = 0$ is not an eigenvalue. Suppose that $\lambda = \alpha^2 > 0$, so

$$y(x) = A \cos \alpha x + B \sin \alpha x.$$

Then the conditions $y(-\pi) = y(\pi) = 0$ yield

$$A \cos \alpha\pi + B \sin \alpha\pi = 0,$$
$$A \cos \alpha\pi - B \sin \alpha\pi = 0.$$

It follows that

$$A \cos \alpha\pi = 0 = B \sin \alpha\pi.$$

Hence either $A = 0$ and $B \neq 0$ with $\alpha\pi$ an even multiple of $\pi/2$, or $A \neq 0$ and $B = 0$ with $\alpha\pi$ an odd multiple of $\pi/2$. Thus the eigenvalues are $\{n^2\}$, and the nth eigenfunction is $y_n(x) = \cos nx/2$ if n is odd, $y_n(x) = \sin nx/2$ if n is even.

4. Just as in Problem 2, $\lambda_0 = 0$ is an eigenvalue with associated eigenfunction $y_0(x) = 1$. If $\lambda = \alpha^2 > 0$ and

$$y(x) = A \cos \alpha x + B \sin \alpha x,$$

then the equations

$$y'(-\pi) = \alpha(A \sin \alpha\pi + B \cos \alpha\pi) = 0,$$
$$y'(\pi) = \alpha(-A \sin \alpha\pi + B \cos \alpha\pi) = 0$$

yield $A \sin \alpha\pi = B \cos \alpha\pi = 0$. If $A = 0$ and $B \neq 0$, then $\cos \alpha\pi = 0$ so $\alpha\pi$ must be an odd multiple of $\pi/2$. If $A \neq 0$ and $B = 0$, then $\sin \alpha\pi = 0$ so $\alpha\pi$ must be an even multiple of $\pi/2$. Therefore the positive eigenvalues are $\{n^2/4\}$ with associated eigenfunctions $y_n(x) = \cos nx/2$ if the integer n is even, $y_n(x) = \sin nx/2$ if n is odd.

5.   If $\lambda = \alpha^2 > 0$ and

$$y(x) = A \cos \alpha x + B \sin \alpha x,$$
$$y'(x) = - A\alpha \sin \alpha x + B\alpha \cos \alpha x$$

then the conditions $y(-2) = y'(2) = 0$ yield

$$A \cos 2\alpha - B \sin 2\alpha = 0,$$
$$- A \sin 2\alpha + B \cos 2\alpha = 0.$$

It follows either that $A = B$ and $\cos 2\alpha = \sin 2\alpha$, or that $A = - B$ and $\cos 2\alpha = - \sin 2\alpha$. The former occurs if

$$2\alpha = \pi/4, 5\pi/4, 9\pi/4, \cdots,$$

the latter if

$$2\alpha = 3\pi/4, 7\pi/4, 11\pi/4, \cdots.$$

Hence the nth eigenvalue is

$$\lambda_n = \alpha_n^2 = (2n -1)^2\pi^2/64$$

for $n = 1, 2, 3, \cdots$, and the associated eigenfunction is

$$y_n(x) = \cos \alpha_n x + \sin \alpha_n x \quad (n \text{ odd})$$

or

$$y_n(x) = \cos \alpha_n x - \sin \alpha_n x \quad (n \text{ even}).$$

6.   (a) If $\lambda = 0$ and $y(x) = A + Bx$, then $y'(0) = B = 0$, so $y(x) = A$. But then $y(1) + y'(1) = A = 0$ also, so $\lambda = 0$ is not an eigenvalue.

(b) If $\lambda = \alpha^2 > 0$ and

$$y(x) = A \cos \alpha x + B \sin \alpha x,$$

then

$$y'(x) = \alpha(- A \sin \alpha x + B \cos \alpha x),$$

so $y'(0) = B\alpha = 0$. Hence $B = 0$ so $y(x) = A \cos \alpha x$. Then

$$y(1) + y'(1) = A(\cos \alpha - \alpha \sin \alpha) = 0 \ ,$$

so α must be a positive root of the equation $\tan \alpha = 1/\alpha$.

7.   (a) If $\lambda = 0$ and $y(x) = A + Bx$, then $y(0) = A = 0$, so $y(x) = Bx$. But then $y(1) + y'(1) = 2B = 0$, so $A = B = 0$ and $\lambda = 0$ is not an eigenvalue.

(b) If $\lambda = \alpha^2 > 0$ and $y(x) = A \cos \alpha x + B \sin \alpha x$, then $y(0) = A = 0$ so $y(x) = B \sin \alpha x$. Hence

$$y(1) + y'(1) = B(\sin \alpha + \alpha \cos \alpha) = 0.$$

so α must be a positive root of the equation $\tan \alpha = - \alpha$.

8.   (a) If $\lambda = 0$ and $y(x) = A + Bx$, then $y(0) = A = 0$, so $y(x) = Bx$. But then $y(1) = y'(1)$ says only that $B = B$. Hence $\lambda_0 = 0$ is an eigenvalue with associated eigenfunction $y_0(x) = x$.

(b) If $\lambda = \alpha^2 > 0$ and $y(x) = A \cos \alpha x + B \sin \alpha x$, then $y(0) = A = 0$ so $y(x) = B \sin \alpha x$. Then $y(1) = y'(1)$ says that $B \sin \alpha = B \alpha \cos \alpha$, so α must be a positive root of the equation $\tan \alpha = \alpha$.

Section 2.10   99

9.   If $y'' + \lambda y = 0$ and $\lambda = -\alpha^2 < 0$,   then

$$y(x) = Ae^{\alpha x} + Be^{-\alpha x}.$$

Then $y(0) = A + B = 0$,   so $B = -A$ and therefore

$$y(x) = A(e^{\alpha x} - e^{-\alpha x}).$$

Hence

$$y'(L) = A\alpha(e^{\alpha L} + e^{-\alpha L}) = 0.$$

But $\alpha \neq 0$ and $e^{\alpha L} + e^{-\alpha L} > 0$,   so $A = 0$.   Thus $\lambda = -\alpha^2$
is not an eigenvalue.

12.   (a) If $\lambda = 0$ and $y(x) = A + Bx$,   then $y(-\pi) = y(\pi)$
means that $A + B\pi = A - B\pi$,   so $B = 0$ and $y(x) = A$.   But
then $y'(-\pi) = y'(\pi)$ implies nothing about   A.   Hence $\lambda_0 = 0$ is an eigenvalue with $y_0(x) = 1$.

(b) If $\lambda = -\alpha^2 < 0$ and

$$y(x) = Ae^{\alpha x} + Be^{-\alpha x},$$

then the conditions $y(-\pi) = y(\pi)$ and $y'(-\pi) = y'(\pi)$
yield the equations

$$Ae^{\alpha\pi} + Be^{-\alpha\pi} = Ae^{-\alpha\pi} + Be^{\alpha\pi},$$
$$Ae^{\alpha\pi} - Be^{-\alpha\pi} = Ae^{-\alpha\pi} - Be^{\alpha\pi}.$$

Addition of these equations yields $2Ae^{\alpha\pi} = 2Ae^{-\alpha\pi}$.   Since
$e^{\alpha\pi} \neq e^{-\alpha\pi}$ because $\alpha \neq 0$,   it follows that $A = 0$.
Similarly $B = 0$.   Thus there are no negative eigenvalues.

(c) If $\lambda = \alpha^2 > 0$ and

$$y(x) = A \cos \alpha x + B \sin \alpha x,$$

then the endpoint conditions yield the equations

$$A \cos \alpha\pi + B \sin \alpha\pi = A \cos \alpha\pi - B \sin \alpha\pi,$$
$$- A \sin \alpha\pi + B \cos \alpha\pi = A \sin \alpha\pi + B \cos \alpha\pi.$$

The first equation implies that $B \sin \alpha\pi = 0$, the second that $A \sin \alpha\pi = 0$. If A and B are not both zero, then it follows that $\sin \alpha\pi = 0$, so $\alpha = n$, an integer. In this case A and B are both arbitrary. Thus $\cos nx$ and $\sin nx$ are two different eigenfunctions associated with the eigenvalue n^2.

13. (a) With $\lambda = 1$, the general solution of $y'' + 2y' + y = 0$ is

$$y(x) = Ae^{-x} + Bxe^{-x}.$$

But then $y(0) = A = 0$ and $y(1) = e^{-1}(A + B) = 0$. Hence $\lambda = 1$ is not an eigenvalue.

(b) If $\lambda < 1$, then the equation $y'' + 2y' + \lambda y = 0$ has characteristic equation $r^2 + 2r + \lambda = 0$. This equation has distinct real roots $[2 \pm (4 - 4\lambda)^{1/2}]/2$; call them r and s. Then the general solution is

$$y(x) = Ae^{rx} + Be^{sx},$$

and the conditions $y(0) = y(1) = 0$ yield the equations

$$A + B = 0, \qquad Ae^r + Be^s = 0.$$

If $A, B \neq 0$, then it follows that $e^r = e^s$. But $r \neq s$, so there is no eigenvalue $\lambda < 1$.

(c) If $\lambda > 1$ let $\lambda - 1 = \alpha^2$, $\lambda = 1 + \alpha^2$. Then the characteristic equation

$$r^2 + 2r + \lambda = (r + 1)^2 + \alpha^2 = 0$$

has roots $-1 \pm \alpha i$, so

$$y(x) = e^{-x}(A \cos \alpha x + B \sin \alpha x).$$

Now $y(0) = A = 0$, so $y(x) = Ae^{-x}\sin \alpha x$. Next, $y(1) = Ae^{-1}\sin \alpha = 0$, so $\alpha = n\pi$ with n an integer. Thus the nth positive eigenvalue is $\lambda_n = n^2\pi^2 + 1$. Because $\lambda = \alpha^2 + 1$, the eigenfunction associated with λ_n is

$$y_n(x) = e^{-x}\sin n\pi x.$$

14. If $\lambda = 1 + \alpha^2$ we first impose the condition $y(0) = 0$ on the solution

$$y(x) = e^{-x}(A \cos \alpha x + B \sin \alpha x)$$

found in Problem 13, and find that $A = 0$. Hence

$$y(x) = Be^{-x}\sin \alpha x,$$
$$y'(x) = B(-e^{-x}\sin \alpha x + e^{-x}\alpha \cos \alpha x),$$

so the condition $y'(1) = 0$ yields $-\sin \alpha + \alpha \cos \alpha = 0$, that is, $\tan \alpha = \alpha$.

15. With $\lambda = \alpha^2 > 0$, the general solution of $x^2y'' + xy' + \alpha^2y = 0$ is
$$y(x) = A \cos(\alpha \ln x) + B \sin(\alpha \ln x).$$

Then $y(1) = A = 0$, so $y(x) = B \sin(\alpha \ln x)$. Now $y(e) = B \sin \alpha = 0$, so $\alpha = n\pi$ with n an integer. Hence the nth positive eigenvalue is $\lambda_n = n^2\pi^2$ with associated eigenfunction $y_n(x) = \sin(n\pi \ln x)$.

16. The characteristic equation of the given Euler-Cauchy equation is $r^2 - 4r + \lambda = 0$, that is,

$$(r^2 - 4r + 4) + \alpha^2 = 0$$

with $\lambda = 4 + \alpha^2$. Then $r = 2 \pm \alpha i$, so the general solution is

$$y(x) = x^2[A \cos(\alpha \ln x) + B \sin(\alpha \ln x)].$$

The condition $y(1) = 0$ yields $A = 0$, so

$$y(x) = Bx^2\sin(\alpha \ln x).$$

Then

$$y(e) = Be^2\sin \alpha = 0,$$

so it follows that α is an integral multiple of π, $\alpha_n = n\pi$, and that the eigenfunction associated with $\lambda_n = 4 + n^2\pi^2$ is $y_n(x) = x^2\sin(n\pi \ln x)$.

17. (a) The endpoint conditions are

$$y(0) = y'(0) = Y''(L) = y^{(3)}(L) = 0.$$

With these conditions, four successive integrations as in Example 5 yield the indicated shape function $y(x)$.

(b) The maximum value y_{max} of $y(x)$ on the closed interval $[0,L]$ must occur either at an interior point where $y'(x) = 0$ or at one of the endpoints $x = 0$ and $x = L$. Now

$$\begin{aligned} y'(x) &= k(4x^3 - 12Lx^2 + 12L^2x) \\ &= 4kx(x^2 - 3Lx + 3L^2) \end{aligned}$$

where $k = w/24EI$, and the quadratic factor has no real zero. Hence $x = 0$ is the only zero of $y'(x)$. But $y(0) = 0$, so it follows that $y_{max} = y(L)$.

18. (a) The endpoint conditions are $y(0) = y'(0) = 0$ and $y(L) = y'(L) = 0$.

(b) The derivative

$$y'(x) = k(4x^3 - 6Lx^2 + 2L^2x)$$

$$= 2kx(2x - L)(x - L)$$

vanishes at $x = 0, L/2, L$. Because $y(0) = Y(L) = 0$, the argument of Problem 17(b) implies that $y_{max} = y(L/2)$.

20. (a) The endpoint conditions are $y(0) = y'(0) = 0$ and $y(L) = y''(L) = 0$.

(b) The only zero of the derivative

$$y'(x) = 2kx(8x^2 - 15Lx + 6L^2)$$

in the interval $[0,L]$ is $x_m = (15 - \sqrt{33})L/16$, and $y(0) = y(L) = 0$, so it follows by the argument of Problem 17(b) that $y_{max} = y(x_m)$.

CHAPTER 3

POWER SERIES SOLUTIONS OF LINEAR EQUATIONS

SECTION 3.1

INTRODUCTION AND REVIEW OF POWER SERIES

The power series method consists of substituting a series $\sum c_n x^n$ into a given differential equation so as to determine what the coefficients $\{c_n\}$ must be in order that the power series will satisfy the equation. It might be pointed out that, if we find a recurrence formula in the form $c_{n+1} = \varphi(n) c_n$, then we can determine the radius of convergence ρ of the series solution directly from the recurrence formula:

$$\rho = \lim_{n \longrightarrow \infty} \left| \frac{c}{c_{n+1}} \right| = \lim_{n \longrightarrow \infty} \left| \frac{1}{\varphi(n)} \right|$$

1. $c_{n+1} = c_n/(n + 1)$; it follows that $c_n = c_0/n!$; $\rho = \infty$

$$y = c_0(1 + x + x^2/2! + x^3/3! + \cdots) = c_0 e^x$$

2. $c_{n+1} = 4c_n/(n + 1)$; $\rho = \infty$

$$y = c_0(1 + 4x + 4^2x^2/2! + 4^3x^3/3! + \cdots) = c_0 e^{4x}$$

3. $c_{n+1} = -3c_n/2(n + 1)$

It follows that $c_n = (-1)^n 3^n c_0/2^n (n!)$; $\rho = \infty$

$$y = c_0(1 - 3x/2 + 3^2x^2/2!2^2 - 3^3x^3/3!2^3$$
$$+ 3^4x^4/4!2^4 - \cdots) = c_0e^{-3x/2}$$

4.   When we substitute $y = \sum c_n x^n$ into the equation $y' + 2xy = 0$, we find that

$$c_1 + \sum [(n+2)c_{n+2} + 2c_n]x^{n+1} = 0.$$

Hence $c_1 = 0$ and $c_{n+2} = -2c_n/(n+2)$. It follows that $c_{odd} = 0$ and $c_{2k} = (-1)^k c_0/k!$, so the solution is

$$y = c_0(1 - x^2 + x^4/2! - x^6/3! + \cdots) = c_0\exp(-x^2).$$

5. $c_{n+3} = c_n/(n+3)$ and $c_1 = c_2 = 0$. It follows that
$c_{3k+1} = c_{3k+2} = 0$ and $c_{3k} = c_0/(3 \cdot 6 \cdots (3k)) = c_0/3^k(k!)$.
$y = c_0(1 + x^3/3 + x^6/2!3^2 + x^9/3!3^3 + \cdots)$
   $= c_0\exp(x^3/3);$ $\rho = \infty$

6. $c_{n+1} = c_n/2 ;$ $\rho = 2$
$y = c_0(1 + x/2 + x^2/2^2 + x^3/2^3 + \cdots) = 2c_0/(2-x)$

7. $c_{n+1} = 2c_n$, so $c_n = 2^n c_0;$ $\rho = 1/2$
$y = c_0(1 + 2x + 4x^2 + 8x^3 + \cdots) = c_0/(1-2x)$

8. $c_{n+1} = -(2n-1)c_n/(2n+2) ;$ $\rho = 1$
$y = c_0(1 + x/2 - x^2/8 + x^3/16 + \cdots) = c_0(1+x)^{1/2}$

9. $c_{n+1} = (n + 2)c_n/(n + 1)$; it follows that $c_n = (n + 1)c_0$

$y = c_0(1 + 2x + 3x^2 + 4x^3 + \cdots) = c_0(1 - x)^{-2}$; $\rho = 1$

10. $c_{n+1} = (2n - 3)c_n/(2n + 2)$; $\rho = 1$

$y = c_0(1 - 3x/2 + 3x^2/8 + x^3/16 + \cdots) = c_0(1 - x)^{3/2}$

11. $c_{n+2} = c_n/(n + 1)(n + 2)$; it follows that

$c_{2k} = c_0/(2k)!$ and $c_{2k+1} = c_1/(2k + 1)!$

$y = c_0(1 + x^2/2! + x^4/4! + x^6/6! + \cdots)$

$\qquad + c_1(x + x^3/3! + x^5/5! + x^7/7! + \cdots)$

$\quad = c_0\cosh x + c_1\sinh x$

12. $c_{n+2} = 4c_n/(n + 1)(n + 2)$

$y = c_0(1 + 2^2x^2/2! + 2^4x^4/4! + \cdots)$

$\qquad + (c_1/2)(2x + 2^3x^3/3! + 2^5x^5/5! + \cdots)$

$\quad = c_0\cosh 2x + (c_1/2)\sinh 2x$

13. $c_{n+2} = - 9c_n/(n + 1)(n + 2)$

$y = c_0(1 - 3^2x^2/2! + 3^4x^4/4! - 3^6x^6/6! + \cdots)$

$\qquad + (c_1/3)(3x - 3^3x^3/3! + 3^5x^5/5! - 3^7x^7/7! + \cdots)$

$\quad = c_0\cos 3x + (c_1/3)\sin 3x$

14. When we substitute $y = \Sigma c_n x^n$ into $y'' + y - x = 0$ and split off the terms of degrees zero and one, we get

$(2c_2 + c_0) + (6c_3 + c_1 - 1)$

$\qquad + \Sigma [(n + 1)(n + 2)c_{n+2} + c_n]x^n = 0.$

Hence $c_2 = -c_0/2$, $c_3 = -(c_1 - 1)/3!$ and $c_{n+2} = -c_n/(n + 1)(n + 2)$. It follows that

$$
\begin{aligned}
y &= x + c_0(1 - x^2/2! + x^4/4! - \cdots) \\
&\quad + (c_1 - 1)(x - x^3/3! + x^5/5! - \cdots) \\
&= x + c_0\cos x + (c_1 - 1)\sin x
\end{aligned}
$$

15. We find that $(n + 1)c_n = 0$ for all $n \geq 0$, so $c_n = 0$ for all $n \geq 0$.

16. When we substitute $y = \sum c_n x^n$ into the differential equation $2xy' = y$, we find that $2nc_n = c_n$ for $n \geq 0$. This can be so only if $c_n = 0$ for all n.

17. We find that $c_0 = c_1 = 0$ and that $c_{n+1} = nc_n$ for $n \geq 1$; it follows that $c_n = 0$ for all $n \geq 0$.

18. When we substitute $y = \sum c_n x^n$ into $x^3 y' = 2y$ we find that $c_0 = c_1 = c_2 = 0$ and that $c_{n+2} = nc_n/2$ for $n \geq 1$. Hence $c_n = 0$ for all n.

19. $c_n = -2^2 c_{n-2}/[n(n - 1)]$ for $n \geq 2$; $c_0 = 0$ and $c_1 = 3$; $y = (3/2)\sin 2x$

20. $c_{n+2} = 4c_n/(n + 1)(n + 2)$; $c_0 = 2$, $c_1 = 0$

$$y = 2(1 + 2^2 x^2/2! + 2^4 x^4/4! + \cdots) = 2\cosh 2x$$

21. $c_{n+1} = (2nc_n - c_{n-1})/[n(n + 1)]$ for $n \geq 1$; $c_0 = 0$ and

$c_1 = 1;$ $y = xe^x$

22. $c_{n+2} = - [(n + 1)c_{n+1} - 2c_n]/(n + 1)(n + 2);$ $c_0 = 1,$ $c_1 = -2$

$y = 1 - 2x + 2x^2 - 4x^3/3 + 2x^4/3 + \cdots$

$= 1 - 2x + 2^2 x^2/2! - 2^3 x^3/3! + 2^4 x^4/4! - \cdots = e^{-2x}$

23. $c_0 = c_1 = 0$ and the recursion relation

$$(n^2 - n + 1)c_n + (n - 1)c_{n-1} = 0$$

for $n \geq 2$ imply that $c_n = 0$ for $n \geq = 0.$

24.   (a) The fact that $y = (1 + x)^\alpha$ satisfies the differential
equation $(1 + x)y' = \alpha y$ follows immediately from the fact
that $y' = \alpha(1 + x)^{\alpha-1}.$

(b) When we substitute $y = \sum c_n x^n$ into the differential
equation $(1 + x)y' = \alpha y$ we get the recurrence formula

$$c_{n+1} = (\alpha - n)c_n/(n + 1).$$

Since $c_0 = 1$ because of the initial condition $y(0) = 1,$
the binomial series in (12) follows.

(c) The function $(1 + x)^\alpha$ and the binomial series must
agree on $(-1, 1)$ because of the uniqueness of solutions of
linear initial value problems.

26.   (b)   The roots of the characteristic equation $r^3 = 1$
are $r_1 = 1,$ $r_2 = \alpha = (- 1 + i\sqrt{3})/2,$ and $r_3 = \beta =$
$(- 1 - i\sqrt{3})/2.$ Then the general solution is

$$y(x) = Ae^x + Be^{\alpha x} + Ce^{\beta x}. \tag{*}$$

Imposing the initial conditions, we get the equations

$$A + B + C = 1$$
$$A + \alpha B + \beta C = 1$$
$$A + \alpha^2 B + \beta^2 C = -1.$$

The solution of this system is

$$A = 1/3, \quad B = (1 - i\sqrt{3})/3, \quad C = (1 + i\sqrt{3})/3.$$

Substitution of these coefficients in (*) and use of Euler's relation $e^{i\theta} = \cos \theta + i \sin \theta$ finally yields the desired result.

SECTION 3.2

SERIES SOLUTIONS NEAR ORDINARY POINTS

Instead of deriving in detail the recurrence relations and solution series for Problems 1 through 15, we indicate where some of these problems and answers originally came from. Each of the differential equations in Problems 1-10 is of the form

$$(Ax^2 + B)y'' + Cxy' + Dy = 0$$

with selected values of the constants A, B, C, D. When we substitute $y = \sum c_n x^n$, shift indices where appropriate, and collect coefficients, we get

$$\sum [An(n - 1)c_n + B(n + 1)(n + 2)c_{n+2} + Cnc_n + Dc_n]x^n = 0.$$

Thus the recurrence relation is

$$c_{n+2} = -\frac{An + (C - A)n + D}{B(n + 1)(n + 2)} c_n \qquad \text{for } n \geq 0.$$

It yields a solution of the form

$$y = c_0 y_{\text{even}} + c_1 y_{\text{odd}}$$

where y_{even} and y_{odd} denote series with terms of even and odd degrees, respectively. The even series $c_0 + c_2 x^2 + c_4 x^4 + \cdots$ converges (by the ratio test) provided that

$$\lim |c_{n+2} x^{n+2} / c_n x^n| = |Ax^2/B| < 1 .$$

Hence its radius of convergence is $\rho = \sqrt{|B/A|}$, as is that of the odd degree series.

1. $c_{n+2} = c_n$

$$y = c_0 \sum_{n=0}^{\infty} x^{2n} + c_1 \sum_{n=0}^{\infty} x^{2n+1} = (c_0 + c_1 x)/(1 - x^2)$$

2. $c_{n+2} = -\frac{1}{2} c_n$

$$y = c_0 \sum_{n=0}^{\infty} (-1)^n \frac{x^{2n}}{2^n} + c_1 \sum_{n=0}^{\infty} (-1)^n \frac{x^{2n+1}}{2^n}$$

3. $c_{n+2} = -\frac{c_n}{(n + 2)}$

$$y = c_0 \sum_{n=0}^{\infty} (-1)^n \frac{x^{2n}}{n!2^n} + c_1 \sum_{n=0}^{\infty} (-1)^n \frac{x^{2n+1}}{(2n + 1)!!}$$

4. $c_{n+2} = -\frac{n + 4}{n + 2} c_n$

$$y = c_0 \sum_{n=0}^{\infty} (-1)^n (n + 1) x^{2n} + \frac{1}{3} c_1 \sum_{n=0}^{\infty} (-1)^n (2n + 3) x^{2n+1}$$

5. $c_{n+2} = -\frac{nc_n}{3(n + 2)}$

$$y = c_0 + c_1 \sum_{n=0}^{\infty} \frac{x^{2n+1}}{(2n + 1)3^n}$$

6. $c_{n+2} = \dfrac{(n - 3)(n - 4)}{(n + 1)(n + 2)} c_n$

   $y = c_0(1 + 6x^2 + x^4) + c_1(1 + x)$

7. $c_{n+2} = -\dfrac{(n - 4)^2}{3(n + 1)(n + 3)} c_n$

   $y = c_0 \left(1 - \dfrac{8}{3} x^2 + \dfrac{8}{27} x^4\right) +$

   $\qquad c_1\left[x - \dfrac{1}{2} x^3 + \dfrac{1}{120} x^5 + 9 \displaystyle\sum_{n=3}^{\infty} \dfrac{(2n - 5)!!]^2(-1)^n}{(2n + 1)! \; 3^n} x^{2n+1}\right]$

8. $c_{n+2} = \dfrac{(n - 4)(n + 4)}{2(n + 1)(n + 2)} c_n$

   $y = c_0 \left(1 - 4x^2 + 2x^4\right) +$

   $\qquad c_1\left[x - \dfrac{5}{4} x^3 + \dfrac{7}{32} x^5 + \displaystyle\sum_{n=3}^{\infty} \dfrac{(2n - 5)!!(2n + 3)!!}{(2n + 1)! \; 2^n} x^{2n+1}\right]$

9. $c_{n+2} = \dfrac{(n + 3)(n + 4)}{(n + 1)(n + 2)} c_n$

   $y = c_0 \displaystyle\sum_{n=0}^{\infty} (n + 1)(2n + 1)x^{2n} + \dfrac{1}{3} c_1 \displaystyle\sum_{n=0}^{\infty} (n + 1)(2n + 3)x^{2n+1}$

10. $c_{n+2} = -\dfrac{(n - 4)}{3(n + 1)(n + 2)} c_n$

   $y = c_0 \left(1 + \dfrac{2}{3} x^2 + \dfrac{1}{27} x^4\right) +$

   $\qquad c_1\left[x + \dfrac{1}{6} x^3 + \dfrac{1}{360} x^5 + 3 \displaystyle\sum_{n=3}^{\infty} \dfrac{(-1)^n(2n-5)!!}{(2n + 1)! \; 3^n} x^{2n+1}\right]$

11. $c_{n+2} = -\dfrac{2(n - 5)}{5(n + 1)(n + 2)} c_n$

   $y = c_1 \left(x - \dfrac{4}{15} x^3 + \dfrac{4}{375} x^5\right) +$

$$c_0 \left[1 - x^2 + \frac{1}{10} x^4 + \frac{1}{750} x^6 + 15 \sum_{n=4}^{\infty} \frac{(2n-7)!! \; 2^n}{(2n)! \; 5^n} x^{2n} \right]$$

12. $c_2 = 0; \quad c_{n+3} = \dfrac{c_n}{(n+2)}$

$$y = c_0 \left[1 + \sum_{n=1}^{\infty} \frac{x^{3n}}{2 \cdot 5 \cdots (3n-1)} \right] + c_1 \sum_{n=0}^{\infty} \frac{x^{3n+1}}{n! \, 3^n}$$

13. $c_2 = 0; \quad c_{n+3} = -\dfrac{c_n}{n+3}$

$$y = c_0 \sum_{n=0}^{\infty} \frac{(-1)^n x^{3n}}{n! \, 3^n} + c_1 \sum_{n=0}^{\infty} \frac{(-1)^n x^{3n+1}}{1 \cdot 4 \cdots (3n+1)}$$

14. $c_2 = 0; \quad c_{n+3} = -\dfrac{c_n}{(n+2)(n+3)}$

$$y = c_0 \left[1 + \sum_{n=1}^{\infty} \frac{(-1)^n x^{3n}}{3^n \cdot n! \cdot 2 \cdot 5 \cdots (3n-1)} \right]$$

$$+ c_1 \sum_{n=0}^{\infty} \frac{(-1)^n x^{3n+1}}{3^n \cdot n! \cdot 1 \cdot 4 \cdots (3n+1)}$$

15. $c_2 = c_3 = 0; \quad c_{n+4} = -\dfrac{c_n}{(n+3)(n+4)}$

$$y = c_0 \left[1 + \sum_{n=1}^{\infty} \frac{(-1)^n x^{4n}}{4^n \cdot n! \cdot 3 \cdot 7 \cdots (4n-1)} \right]$$

$$+ c_1 \sum_{n=0}^{\infty} \frac{(-1)^n x^{4n+1}}{4^n \cdot n! \cdot 5 \cdot 9 \cdots (4n+1)}$$

16. The recurrence formula is

$$c_{n+2} = -(n-1)c_n/(n+1)$$

for $n \geq 0$. This yields $c_3 = c_5 = \cdots = 0$ and
$c_{2n} = (-1)^{n-1} c_0/(2n-1)$ for $n \geq 1$. Hence

$$y = c_1 x + c_0 (1 + x^2 - x^4/3 + x^6/5 - \cdots)$$

$$= c_1 x + c_0 (1 + x \tan^{-1}x).$$

With $c_0 = y(0) = 0$ and $c_1 = y'(0) = 1$ we obtain the desired particular solution $y(x) = x$.

17. The recurrence relation

$$c_{n+2} = -(n - 2)c_n/(n + 1)(n + 2)$$

yields $c_2 = c_0 = y(0) = 1$ and $c_4 = c_6 = \cdots = 0$. Because $c_1 = y'(0) = 0$, it follows also that $c_1 = c_3 = c_5 = \cdots = 0$. Thus the desired particular solution is $y(x) = 1 + x^2$.

18. The substitution $t = x - 1$ yields $y'' + ty' + y = 0$, where primes now denote differentiation with respect to t. When we substitute $y = \sum c_n t^n$ we get the recurrence relation

$$c_{n+2} = - c_n/(n + 2)$$

for $n \geq 0$. The initial conditions give $c_0 = 2$ and $c_1 = 0$, so $c_{odd} = 0$ and it follows that

$$y = 2(1 - t^2/2 + t^4/2 \cdot 4 - t^6/2 \cdot 4 \cdot 6 + \cdots)$$

$$= 2[1 - (x-1)^2/2 + (x-1)^4/2 \cdot 4 - (x-1)^6/2 \cdot 4 \cdot 6 + \cdots]$$

19. $y = (1/3)\sum(2n + 3)(x - 1)^{2n+1}$; converges if $0 < x < 2$.

20. The substitution $t = x - 3$ yields

$$(t^2 + 1)y'' - 4ty' + 6y = 0 ,$$

where primes now denote differentiation with respect to t. When we substitute $y = \sum c_n t^n$ we get the recurrence relation

$$c_{n+2} = - (n - 2)(n - 3)c_n/(n + 1)(n + 2)$$

for $n \geq 0$. The initial conditions give $c_0 = 2$ and $c_1 = 0$. It follows that $c_{odd} = 0$, $c_2 = -6$ and $c_4 = c_6 = \cdots = 0$, so the solution reduces to

$$y = 2 - 6t^2 = 2 - 6(x - 3)^2. \char94 0$$

21. $y = 1 + 4(x + 2)^2$; converges for all x.

22. The differential equation we wish to solve is

$$(x^2 + 6x)y'' + (3x + 9)y' - 3y = 0.$$

The substitution $t = x + 3$ yields

$$(t^2 + 9)y'' + 3ty' - 3y = 0,$$

with primes now denoting differentiation with respect to t. When we substitute $y = \sum c_n t^n$ we get the recurrence relation

$$c_{n+2} = (n + 3)(n - 1)c_n/9(n + 1)(n + 2)$$

for $n \geq 0$. The initial conditions give $c_0 = 0$ and $c_1 = 2$. It follows that $c_{even} = 0$ and $c_3 = c_5 = \cdots = 0$, so

$$y = 2t = 2x + 6.$$

23. $2c_2 + c_0 = 0$; $(n + 1)(n + 2)c_{n+2} + c_n + c_{n-1} = 0$ for $n \geq 1$

$$y_1 = 1 - (1/2)x^2 - (1/6)x^3 + \cdots;$$

$$y_2 = x - (1/6)x^3 - (1/12)x^4 + \cdots$$

24. Substitution of $y = \sum c_n x^n$ yields

$$2c_2 + \sum [2c_n + (n + 1)(n + 2)c_{n+1}$$
$$- (n + 2)(n + 3)c_{n+3}]x^{n+1} = 0 .$$

Thus $c_2 = 0$ and

$$c_{n+3} = [2c_n + (n + 1)(n + 2)c_{n+1}]/(n + 2)(n + 3)$$

for $n \geq 0$. We find that $c_3 = (c_0 + c_1)/3$, $c_4 = c_1/6$, and $c_5 = (c_0 + c_1)/5$, so

$$y = c_0(1 + x^3/3 + x^5/5 + \cdots) + c_1(x + x^3/3 + x^4/6 + \cdots).$$

25. $c_2 = c_3 = 0$; $(n + 3)(n + 4)c_{n+4} + (n + 1)c_{n+1} + c_n = 0$ for $n \geq 0$;

$$y_1 = 1 - (1/12)x^4 + (1/126)x^7 + \cdots ;$$

$$y_2 = x - (1/12)x^4 - (1/20)x^5 + \cdots$$

26. When we substitute $y = \sum c_n x^n$, shift indices and collect coefficients, we find that

$$2c_2 + 6c_3x + 12c_4x^2 + (2c_2 + 20c_5)x^3$$
$$+ \sum [c_n + (n+2)(n+3)c_{n+3} + (n+5)(n+6)c_{n+6}]x^{n+4} = 0.$$

Hence $c_2 = c_3 = c_4 = c_5 = 0$ and

$$c_{n+6} = - [c_n + (n + 2)(n + 3)c_{n+3}]/(n + 5)(n + 6)$$

for $n \geq 0$. It follows that $c_6 = - c_0/30$, $c_7 = - c_1/42$, $c_8 = 0$, $c_9 = c_0/72$, and $c_{10} = c_1/90$, so

$$y = c_0(1 - x^6/30 + x^9/72 + \cdots)$$
$$+ c_1(x - x^7/42 + x^{10}/90 + \cdots).$$

27. $y = 1 - x - (1/2)x^2 + (1/3)x^3 - (1/24)x^4 + (1/30)x^5$

   $+ (29/720)x^6 - (13/630)x^7 - (143/40320)x^8 + \cdots;$

   $y(0.5) \approx 0.4156$

28. When we substitute $y = \sum c_n x^n$ and $e^{-x} = \sum (-1)^n x^n/n!$, and then collect coefficients of the terms involving 1, x, and x^2 we find that

   $$c_2 = - c_0/2, \quad c_3 = (c_0 - c_1)/6, \quad \text{and}$$
   $$c_4 = (- c_0/2 + c_1 - c_2)/12 = c_1/12.$$

   It follows that

   $$y = c_0(1 - x^2/2 + x^3/6 + \cdots) + (x - x^3/6 + x^4/12 + \cdots).$$

29. $y_1 = 1 - (1/2)x^2 + (1/720)x^6 + \cdots$

   $y_2 = x - (1/6)x^3 - (1/60)x^5 + \cdots$

30. When we substitute $y = \sum c_n x^n$ and $\sin x = \sum (-1)^n x^{2n+1}/(2n + 1)!$, and then collect coefficients of the terms involving x, x^2, x^3, and x^4, we find that

   $$c_2 = - c_0, \quad c_3 = - c_1/3,$$
   $$c_4 = (c_0/6 - 2c_2)/12 = 13c_0/72,$$
   $$\text{and } c_5 = (c_1/6 - 2c_3)/20 = c_1/24 .$$

   It follows that

$$y = c_0(1 - x^2 + 13x^4/72 + \cdots)$$
$$+ c_1(x - x^3/3 + x^5/24 + \cdots).$$

SECTION 3.3

REGULAR SINGULAR POINTS

1. Upon division of the given differential equation by x   we
    see that $P(x) = 1 - x^2$ and $Q(x) = (\sin x)/x$.   Because
    both are analytic at $x = 0$ (in particular, $(\sin x)/x \longrightarrow$
    1 as $x \longrightarrow 0$)   it follows that $x = 0$ is an ordinary
    point.

2. Division of the differential equation by   x   yields

    $$y'' + xy' + (1/x)(e^x - 1)y = 0.$$

    Since

    $$(1/x)(e^x - 1) = (1/x)(x + x^2/2! + x^3/3! + \cdots)$$
    $$= 1 + x/2! + x^2/3! + \cdots$$

    is analytic at the origin, $x = 0$ is an ordinary point.

3. When we rewrite the given equation in the standard form of
    Equation (3) in this section, we see that $p(x) = (\cos x)/x$
    and $q(x) = x$.   Because $(\cos x)/x \longrightarrow \infty$ as $x \longrightarrow 0$ it
    follows that $p(x)$ is not analytic, so $x = 0$ is an
    irregular singular point.

4. When we rewrite the given equation in the standard form of
    Equation (3), we have $p(x) = 2/3$ and $q(x) = (1 - x^2)/3x$.
    Since $q(x)$ is not analytic at the origin, $x = 0$ is an
    irregular singular point.

5. In the standard form of Equation (3) we have $p(x) =$
    $2/(1 + x)$ and $q(x) = 3x^2/(1 + x)$.   Both are analytic, so

$x = 0$ is a regular singular point. The indicial equation is

$$r(r - 1) + 2r = r^2 + r = r(r + 1) = 0,$$

so the exponents are $r_1 = 0$ and $r_2 = -1$.

6. In the standard form of Equation (3) we have $p(x) = 2/(1 - x^2)$ and $q(x) = -2/(1 - x^2)$, so $x = 0$ is a regular singular point with $p_0 = 2$ and $q_0 = -2$. The indicial equation is $r^2 + r - 2 = 0$, so the exponents are $r = -2, 1$.

7. In the standard form of Equation (3) we have $p(x) = (6 \sin x)/x$ and $q(x) = 6$, so $x = 0$ is a regular singular point with $p_0 = q_0 = 6$. The indicial equation is $r^2 + 5r + 6 = 0$, so the exponents are $r_1 = -2$ and $r_2 = -3$.

8. In the standard form of Equation (3) we have $p(x) = 21/(6 + 2x)$ and $q(x) = 9(x^2 - 1)/(6 + 2x)$, so $x = 0$ is a regular singular point with $p_0 = 7/2$ and $q_0 = -3/2$. The indicial equation simplifies to $2r^2 + 5r - 3 = 0$, so the exponents are $r = -3, 1/2$.

9. The only singular point of the given equation is $x = 1$. Upon substituting $t = x - 1$, $x = t + 1$ we get the transformed equation

$$- ty'' + (t + 1)y' + (t + 1)^2 y = 0$$

where primes now denote differentiation with respect to t. In the standard form of Equation (3) we have $p(t) = -1 - t$ and $q(t) = -t(1 + t)^2$. Both are analytic, so it follows that $x = 1$ is a regular singular point of the original equation.

10. Regular singular point x = 1

11. Regular singular points x = + 1   and   x = - 1

12. Irregular singular point x = 2

13. The singular points of the given equation are x = +2   and
    x = -2.

    x = +2: The substitution t = x - 2,   x = t + 2   yields the
    transformed equation

    $$t(t + 4)y" + ty' + (t + 4)y = 0.$$

    In the standard form of Equation (3) we have p(t) =
    t/(4 + t)   and   q(t) = t. Both are analytic at t = 0,   so
    x = +2   is a regular singular point of the original
    equation.

    x = -2: The substitution t = x + 2,   x = t - 2   yields the
    transformed equation

    $$t(t - 4)y" + (t - 4)y' + ty = 0.$$

    Now p(t) = 1   and   q(t) = $t^2/(t - 4)$. Both are analytic at
    t = 0,   so x = -2   is also a regular singular point of the
    original equation.

14. Irregular singular points x = -3, 3

15. Regular singular point x = 2

16. Irregular singular point x = 0   and regular singular point
    x = 1

Instead of deriving in detail the solutions given below for the
differential equations in Problems 17-31, we indicate here where
some of these equations and solutions came from. Each of the
differential equations in Problems 17-20 is of the form

$$Axy'' + By' + Cy = 0$$

with indicial equation $Ar^2 + (B - A)r = 0$. Substitution of $y = \sum c_n x^{n+r}$ into the differential equation yields the recurrence formula

$$c_n = - C\ c_{n-1}/[A(n + r)^2 + (B - A)(n + r)]$$

for $n \geq 1$. In these problems the exponents $r_1 = 0$ and $r_2 = (A - B)/A$ do not differ by an integer, so this recurrence formula yields two linearly independent Frobenius series solutions when we apply it separately with $r = r_1$ and with $r = r_2$.

The differential equations in Problems 21-24, 27-29, and 31 are all of the form

$$Ax^2y'' + Bxy' + (C + Dx^2)y = 0$$

with indical equation

$$\varphi(r) = Ar^2 + (B - A)r + C = 0.$$

substitution of $y = \sum c_n x^{n+r}$ into the differential equation yields

$$\varphi(r)c_0 x^r + \varphi(r + 1)c_1 x^{r+1} + \sum\ [\varphi(n + r)c_n + Dc_{n-2}]x^{n+r} = 0. \qquad (*)$$

In each of Problems 21-24 the exponents r_1 and r_2 do *not* differ by an integer. Hence when we substitute either $r = r_1$ or $r = r_2$ into Equation (*), we find that c_0 is arbitrary, that $c_1 = 0$, and that

$$c_n = - Dc_{n-2}/[A(n+r)^2 + (B - A)(n + r) + C]$$

for $n \geq 2$. Thus this recurrence formula yields two linearly independent Frobenius series solutions when we apply it separately with $r = r_1$ and with $r = r_2$.

In Problems 27-29 and 31, by contrast, the exponents r_1 and r_2 = r_1 - 1 *do* differ by an integer. Hence when we substitute the smaller exponent $r = r_2$ into Equation (*), we find that c_0 and c_1 are *both* arbitrary, and that c_n is given (for $n \geq 2$) by the recurrence formula above. Thus the smaller exponent yields two linearly independent Frobenius series solutions.

17. $y_1 = \cos \sqrt{x}$, $\quad y_2 = \sin \sqrt{x}$

18. $y_1 = \sum_{n=0}^{\infty} \dfrac{x^n}{n!(2n + 1)!!}$

$y_2 = x^{-1/2} \sum_{n=0}^{\infty} \dfrac{x^n}{n!(2n - 1)!!}$

19. $y_1 = x^{3/2} \left[1 + 3 \sum_{n=1}^{\infty} \dfrac{x^n}{n!(2n + 3)!!} \right]$

$y_2 = 1 - x - \sum_{n=2}^{\infty} \dfrac{x^n}{n!(2n - 3)!!}$

20. $y_1 = x^{1/3} \sum_{n=0}^{\infty} \dfrac{(-1)^n 2^n x^n}{n! \cdot 4 \cdot 7 \cdots (3n + 1)}$

$y_2 = \sum_{n=0}^{\infty} \dfrac{(-1)^n 2^n x^n}{n! \cdot 2 \cdot 5 \cdots (3n - 1)}$

21. $y_1 = x \left[1 + \sum_{n=1}^{\infty} \dfrac{x^{2n}}{n! \cdot 7 \cdot 11 \cdots (4n + 3)} \right]$

$y_2 = x^{-1/2} \sum_{n=0}^{\infty} \dfrac{x^{2n}}{n! \cdot 1 \cdot 5 \cdots (4n + 1)}$

22. $y_1 = x^{3/2} \left[1 + \sum_{n=1}^{\infty} \dfrac{(-1)^n x^{2n}}{n! \cdot 9 \cdot 13 \cdots (4n + 5)} \right]$

$y_2 = x^{-1} \left[1 + \sum_{n=1}^{\infty} \dfrac{(-1)^{n-1} x^{2n}}{n! \cdot 3 \cdot 7 \cdots (4n - 1)} \right]$

23. $y_1 = x^{1/2} \left[1 + \sum\limits_{n=1}^{\infty} \dfrac{x^{2n}}{2^n n! \cdot 19 \cdot 31 \cdots (12n + 7)} \right]$

$y_2 = x^{-2/3} \left[1 + \sum\limits_{n=1}^{\infty} \dfrac{x^{2n}}{2^n n! \cdot 5 \cdot 17 \cdots (12n - 7)} \right]$

24. $y_1 = x^{1/3} \left[1 + \sum\limits_{n=1}^{\infty} \dfrac{(-1)^n x^{2n}}{2^n n! \cdot 7 \cdot 13 \cdots (6n + 1)} \right]$

$y_2 = 1 + \sum\limits_{n=1}^{\infty} \dfrac{(-1)^n x^{2n}}{2^n n! \cdot 5 \cdot 11 \cdots (6n - 1)}$

25. $y_1 = x^{1/2} \sum\limits_{n=0}^{\infty} \dfrac{(-1)^n x^n}{n! 2^n} = x^{1/2} e^{-x/2}$

$y_2 = 1 + \sum\limits_{n=1}^{\infty} \dfrac{(-1)^n x^n}{(2n - 1)!!}$

26. $y_1 = x^{1/2} \sum\limits_{n=0}^{\infty} \dfrac{x^{2n}}{n! 2^n} = x^{1/2} \exp\left(\dfrac{x^2}{2}\right)$

$y_2 = 1 + \sum\limits_{n=1}^{\infty} \dfrac{2^n x^{2n}}{3 \cdot 7 \cdots (4n - 1)}$

27. $y_1 = (\cos 3x)/x, \quad y_2 = (\sin 3x)/x$

28. $y_1 = (\cosh 2x)/x, \quad y_2 = (\sinh 2x)/x$

29. $y_1 = (1/x)\cos(x/2), \quad y_2 = (1/x)\sin(x/2)$

30. $y_1 = \cos x^2, \quad y_2 = \sin x^2$

31. $y_1 = x^{1/2}\cosh x, \quad y_2 = x^{1/2}\sinh x$

32. Exponents $r_1 = 1$ and $r_2 = -1/2$

$c_n = -(n - r - 3)c_{n-1}/[2(n + r)^2 - (n + r) - 1]$

$y_1 = x + 3x^2/5 + 3x^3/35 + x^4/315$ (terminates)

$y_2 = x^{-1/2}(1 - 3x/2 - 3x^2/8 + x^3/48 + \cdots)$

33. $y_1 = x^{-1}(1 + 10x + 5x^2 + 10x^3/9 + \cdots)$

     $y_2 = x^{1/2}(1 + 11x/20 - 11x^2/224 + 671x^3/24192 + \cdots)$

34.   Exponents $r_1 = 1$ and $r_2 = -1/2$

     $y_1 = x(1 - x^2/42 + x^4/1320 + \cdots)$

     $y_2 = x^{-1/2}(1 - 7x^2/24 + 19x^4/3200 + \cdots)$

36.   (a) When we substitute $y = \sum c_n x^{n+r}$ we find that $r = 0$ is the only possible value.

     (b) Now we find that $r = -B/A$ is the only possible value.

     (c) When we substitute $y = \sum c_n x^{n+r}$ with $c_0 \neq 0$ the term involving x^r is $Bc_0 x^r$,   which cannot vanish.

38.   Bessel's equation of order 1/2 is of the same form as the equations in Problems 27-29 and 31 discussed above, with $r_1 = 1/2$ and $r_2 = -1/2$ differing by   1.   Hence the smaller exponent $r_2 = -1/2$ yields two linearly independent Frobenius series solutions.

SECTION 3.4

METHOD OF FROBENIUS -- THE EXCEPTIONAL CASES

Each of the differential equations in Problems 1-6 is of the form

$$xy'' + (A + Bx)y' + Cy = 0.$$

The origin is a regular singular point with exponents $r = 0$ and $r = 1 - A$, so if A is an integer then we have an exceptional case of the method of Frobenius. When we substitute $y = \sum c_n x^{n+r}$ in the differential equation we find that the coefficient of x^{n+r} is

$$[(n + r)^2 + (A - 1)(n + r)]c_n + [B(n + r) + C - B]c_{n-1} = 0. \qquad (*)$$

Case 1: In each of Problems 1-4 we have $A \geq 2$ and $B = C$, so the larger exponent $r_1 = 0$ and the smaller exponent $r_2 = 1 - A = -N$ differ by a positive integer. When we substitute the smaller exponent $r = -N$ in Equation (*) it simplifies to

$$n(n - N)c_n + B(n - N)c_{n-1} = 0. \qquad (1)$$

This equation determines $c_1, c_2, \cdots, c_{N-1}$ in terms of c_0, thereby yielding the solution

$$y_1 = x^{-N}(c_0 + c_1 x + \cdots + c_{N-1}x^{n-1}) .$$

When $n = N$ Equation (1) reduces to

$$0 \cdot c_n + 0 \cdot c_{N-1} = 0,$$

so c_n may be chosen arbitrarily. For $n > N$ Equation (1) yields the recurrence formula $c_n = -Bc_{n-1}/n$, and the second solution is of the form

$$y_2 = c_n + c_{n+1}x + c_{n+2}x^2 + \cdots$$

Case 2: If $A \leq 0$ then the larger exponent $r_1 = 1 - A = N$ and the smaller exponent $r_2 = 0$ again differ by a positive integer. In Problems 5 and 6 we have this case with $B = -1$. When we substitute the smaller exponent $r = 0$ in Equation (*) it simplifies to

$$n(n - N)c_n - (n - C - 1)c_{n-1} = 0. \qquad (2)$$

This equation determines $c_1, c_2, \cdots, c_{N-1}$ in terms of c_0. When $n = N$ it reduces to

$$0 \cdot c_N - (N - C - 1)c_{N-1} = 0. \tag{3}$$

If either $N - C - 1 = 0$ or $c_{N-1} = 0$ (the latter happens in Problem 5) then c_N can be chosen arbitrarily, and finally $c_{N+1}, c_{N+2}, \cdots$ are determined in terms of c_N. Thus we get two Frobenius series solutions

$$y_1 = c_0 + c_1 x + \cdots + c_{N-1}x^{N-1},$$

$$y_2 = c_N x^N + c_{N+1}x^{N+1} + \cdots .$$

On the other hand, if (as in Problem 6) neither $N - C - 1 = 0$ nor $c_{N-1} = 0$, then c_N cannot be chosen so as to satisfy Equation (3), and hence there is no Frobenius series solution corresponding to the smaller exponent $r_2 = 0$. We therefore find the single Frobenius series solution by substituting the larger exponent $r_1 = N$ in Equation (*) and using the resulting recurrence relation to determine $c_1, c_2, c_3, \cdots$ in terms of c_0.

1. $\quad y_1 = x^{-2}(1 + x)$

$\quad\quad y_2 = 1 + 2 \sum\limits_{n=1}^{\infty} \dfrac{x^n}{(n + 2)!}$

2. $\quad y_1 = x^{-4}(1 + x + \frac{1}{2}x^2 + \frac{1}{6}x^3)$

$\quad\quad y_2 = 1 + 24 \sum\limits_{n=1}^{\infty} \dfrac{x^n}{(n + 4)!}$

3. $\quad y_1 = x^{-4}(1 - 3x + \frac{9}{2}x^2 - \frac{9}{2}x^3)$

$\quad\quad y_2 = 1 + 24 \sum\limits_{n=1}^{\infty} \dfrac{(-1)^n 3^n x^n}{(n + 4)!}$

4. $\quad y_1 = x^{-5}(1 - \frac{3}{5} x + \frac{9}{50} x^2 - \frac{9}{250} x^3 + \frac{27}{1000} x^4)$

$y_2 = 1 + 120 \sum_{n=1}^{\infty} \frac{(-1)^n 3^n x^n}{(n + 5)! 5^n}$

5. $\quad y_1 = 1 + \frac{3}{4} x + \frac{1}{4} x^2 + \frac{1}{24} x^3$

$y_2 = x^5 \left[1 + 120 \sum_{n=1}^{\infty} \frac{(n + 1) x^n}{(n + 5)!} \right]$

6. Here $A = -3$, $B = -1$, $C = 1/2$, $r_1 = N = 4$, and $r_2 = 0$, and Equation (2) is

$$n(n - 4)c_n - (n - 3/2)c_{n-1} = 0.$$

Starting with $c_0 = 1$, this equation gives $c_1 = 1/6$, $c_2 = 1/48$, $c_3 = 1/96$. With $n = 4$ it reduces to

$$0 \cdot c_4 - (7/2)(1/96) = 0,$$

so c_4 cannot be chosen. We therefore start over by substituting $r_1 = 4$ in Equation (*) and get the recurrence formula

$$c_n = (2n + 5)c_{n-1}/2n(n + 4)$$

for $n \geq 1$. This yields the single Frobenius series solution

$$y_1 = x^4[1 + (8/5)\sum (2n + 5)!! x^n/2^n n! (n + 4)!].$$

7. $\quad y_1 = x^{-2}(2 - 6x + 9x^2);$ $\quad y_2 = \sum (-1)^{n-1} 3^n x^n/(n + 2)!$

8. The exponents are $r = 0, 4$. When we substitute $y = \sum c_n x^n$ (corresponding to $r = 0$) in the differential equation we get

$$(n - 4)c_n - (n - 3)c_{n-1} = 0$$

for $n \geq 1$. Starting with $c_0 = 1$, we compute $c_1 = 2/3$, $c_2 = 1/3$, and $c_3 = 0$. Hence

$$y_1 = 3 + 2x + x^2$$

is one solution. Because $c_3 = 0$ we can choose $c_4 = 1$. Then our recurrence formula above yields $c_5 = 2$, $c_6 = 3$, $c_7 = 4$, $\cdots$. Hence the second solution is

$$y_2 = x^4(1 + 2x + 3x^2 + 4x^3 + \cdots)$$

$$= x^4/(1 - x)^2,$$

with the closed form coming from the derivative of the geometric series $1/(1 - x) = \sum x^n$.

9. $y_1 = 1 + x^2/2^2 + x^4/2^2 4^2 + x^6/2^2 4^2 6^2 + \cdots ;$

$y_2 = y_1(\ln x - x^2/4 + 5x^4/128 - 23x^6/3456 + \cdots)$

10. $y_1 = x(1 - x^2/2^2 + x^4/2^2 4^2 - x^6/2^2 4^2 6^2 + \cdots)$

$y_2 = y_1 \int x \cdot x^{-2} (1 - x^2/4 + x^4/64 - x^6/2304 + \cdots)^{-2} dx$

$\quad = y_1 \int x^{-1}(1 - x^2/2 + 3x^4/32 - 5x^6/576 + \cdots)^{-1} dx$

$\quad = y_1 \int x^{-1}(1 + x^2/2 + 5x^4/32 + 23x^6/576 + \cdots) dx$

$y_2 = y_1 (\ln x + x^2/4 + 5x^4/128 + 23x^6/3456 + \cdots)$

11. $y_1 = x^2[1 - 2x + (3/2)x^2 - (2/3)x^3 + \cdots]$

$y_2 = y_1[\ln x + 3x + (11/4)x^2 + (49/18)x^3 + \cdots]$

12. $y_1 = x^2 \left[1 - \dfrac{x}{2} + \dfrac{3x^2}{20} - \dfrac{x^3}{30} + \dfrac{x^4}{168} - \cdots \right]$

$y_2 = y_1 \displaystyle\int e^{-x} x^{-4} \left[1 - \dfrac{x}{2} + \dfrac{3x^2}{20} - \dfrac{x^3}{30} + \dfrac{x^4}{168} - \cdots \right]^{-2} dx$

$= y_1 \displaystyle\int x^{-4} e^{-x} \left[1 - x + \dfrac{11x^2}{20} - \dfrac{13x^3}{60} + \dfrac{569x^4}{8400} + \cdots \right]^{-1} dx$

$= y_1 \displaystyle\int x^{-4} \left[1 - \dfrac{x^2}{20} + \dfrac{3x^4}{100} + \cdots \right] dx$

$y_2 = y_1 \left[- \dfrac{1}{3x^3} + \dfrac{1}{20x} + \dfrac{3x}{100} + \cdots \right]$ $\qquad$ (no logarithmic term)

13. $y_1 = x^3 [1 - 2x + 2x^2 - (4/3)x^3 + \cdots]$

$y_2 = y_1 [2 \ln x - 1/2x^2 - 2/x + (4/3)x + \cdots]$

14. $y_1 = x^2 \left[1 - \dfrac{2x}{5} + \dfrac{x^2}{10} - \dfrac{2x^3}{105} + \dfrac{x^4}{336} - \cdots \right]$

$y_2 = y_1 \displaystyle\int x^{-1} e^{-x} (y_1)^{-2} dx$

$= y_1 \displaystyle\int \left[x^{-5} - \dfrac{x^{-4}}{5} - \dfrac{x^{-3}}{50} + \dfrac{13x^{-2}}{1750} + 0 \cdot x^{-1} + \cdots \right] dx$

$y_2 = y_1 \left[- \dfrac{x^{-4}}{4} + \dfrac{x^{-3}}{15} + \dfrac{x^{-2}}{100} - \dfrac{13x^{-1}}{1750} + 0 \cdot \ln x + \cdots \right]$

Thus y_2 contains no logarithmic term.

16. $y_1 = x^{3/2} \left[1 + \displaystyle\sum_{n=1}^{\infty} \dfrac{(-1)^n x^{2n}}{2^n n! \cdot 5 \cdot 7 \cdots (2n + 3)} \right]$

$y_2 = x^{-3/2} \left[1 + \displaystyle\sum_{n=1}^{\infty} \dfrac{(-1)^n x^{2n}}{2^n \cdot n! \cdot (-1) \cdot 1 \cdot 3 \cdots (2n - 3)} \right]$

18. When we substitute

$$y(x) = C\, y_1 \ln x + \sum b_n x^n$$

in the differential equation $xy'' - x = 0$ we find that $b_1 = -b_0 = -C$ and that

$$n(n + 1)b_{n+1} - b_n = -(2n + 1)C/n!\,(n + 1)!$$

for $n \geq 1$. To solve this recursion formula we take $C = 1$ and substitute $b_n = c_n/(n - 1)!\,n!$. The result is

$$c_{n+1} - c_n = -\frac{2n + 1}{n(n + 1)} = -\frac{1}{n} - \frac{1}{n + 1}.$$

Starting with $c_1 = b_1 = -1$, it follows readily by induction on n that $c_n = -(H_n + H_{n+1})$.

BESSEL'S EQUATION

Of course Bessel's equation is the most important special ordinary differential equation in mathematics, and every student should be exposed at least to Bessel functions of the first kind. Though Bessel functions of integral order can be treated without the gamma function, the subsection on the gamma function is needed in Chapter 4 on Laplace transforms. The final subsections on Bessel function identities and the parametric Bessel equation will not be needed until Section 9.4, and therefore may be considered optional at this point in the course.

2. (a) $\Gamma([2n + 1]/2) = ([2n - 1]/2)\,\Gamma([2n - 1]/2)$

$\qquad\qquad = ([2n - 1]/2)\cdot([2n - 3]/2)\,\Gamma([2n - 3]/2)$

$\qquad\qquad = \cdots$

$\qquad\qquad = 2^{-n}(2n - 1)(2n - 3)\,\cdots\,(3)(1)\,\Gamma(1/2)$

$\quad \Gamma([2n + 1]/2) = 2^{-n}(1)(3)\,\cdot\,\cdot\,\cdot\,(2n - 1)\sqrt{\pi}$

(b) $J_{1/2}(x) = \sum_0^\infty \dfrac{(-1)^m \, x^{2m + 1/2}}{m!\,\Gamma(m + 3/2)\,2^{2m + 1/2}}$

$\qquad\qquad = \sqrt{(2/x)} \sum_0^\infty \dfrac{(-1)^m \, x^{2m+1}}{m!\; 2^{-m-1}(2m + 1)!!\; \sqrt{\pi}\; 2^{2m+1}}$

$\qquad\qquad = \sqrt{(2/\pi x)} \sum_0^\infty \dfrac{(-1)^m \, x^{2m+1}}{(2)(4)\cdots(2n)(1)(3)\cdots(2m + 1)}$

$\qquad\qquad = \sqrt{(2/\pi x)} \sum_0^\infty \dfrac{(-1)^m \, x^{2m+1}}{(2m + 1)!}$

$\qquad J_{1/2}(x) = \sqrt{(2/\pi x)} \, \sin x$

4.   With $p = 1/2$ in Equation (26) we have

$J_{3/2}(x) = (1/x)J_{1/2}(x) - J_{-1/2}(x)$

$\qquad\qquad = (1/x)\sqrt{(2/\pi x)}\sin x - \sqrt{(2/\pi x)}\cos x$

$J_{3/2}(x) = \sqrt{(2/\pi x^3)} \, (\sin x - x \cos x)$

5.   Starting with $p = 3$ in Equation (26) we get

$J_4(x) = (6/x)J_3(x) - J_2(x)$

$\qquad\quad = (6/x)[(4/x)J_2(x) - J_1(x)] - J_2(x)$

$\qquad\quad = (24/x^2 - 1)[(2/x)J_1(x) - J_0(x)] - (6/x)J_1(x)$

$J_4(x) = (1/x^2)(24 - x^2)J_0(x) + (8/x^3)(6 - x^2)J_1(x)$

8.   When we carry out the differentiations indicated in Equations (22) and (23), we get

$\qquad p\,x^{p-1}J_p(x) + x^p J_p{}'(x) = x^p J_{p-1}(x),$

$$- p \; x^{-p-1} J_p(x) + x^{-p} J_p{}'(x) = - x^{-p} J_{p+1}(x).$$

When we solve these two equations for $J_p{}'(x)$ we get Equations (24) and (25).

10. When we add equations (24) and (25) we get

$$J_p{}'(x) = (1/2) [J_{p-1}(x) - J_{p+1}(x)],$$

so

$$J_p{}''(x) = (1/2) [J_{p-1}{}'(x) - J_{p+1}{}'(x)].$$

Replacing p with $p - 1$ and with $p + 1$ in the first equation, we get

$$J_{p-1}{}'(x) = (1/2) [J_{p-2}(x) - J_p(x)]$$

and

$$J_{p+1}{}'(x) = (1/2) [J_p(x) - J_{p+2}(x)].$$

When we use these equations to substitute for $J_{p-1}{}'(x)$ and $J_{p+1}(x)$ in the equation for $J_p{}''(x)$ above, we find that

$$J_p{}''(x) = (1/4) [J_{p-2}(x) - 2 \; J_p(x) + J_{p+2}(x)].$$

11. $\Gamma(p + m + 1) = (p + m)(p + m - 1) \cdots (p + 2)(p + 1)\Gamma(p + 1)$

12. Substitution of the power series of Problem 12 yields

$$y(x) = x^2 \cdot \frac{x^{5/2}(A + \cdots) + x^{-5/2}(B + \cdots)}{x^{1/2}(C + \cdots) + x^{-1/2}(D + \cdots)}$$

$$= \frac{x^5 (A + \cdots) + (B + \cdots)}{x(C + \cdots) + (D + \cdots)}$$

where $A = 1/2^{5/2}\Gamma(7/2)$, $B = 1/2^{-5/2}\Gamma(-3/2)$,

$C = 1/2^{1/2}\Gamma(3/2)$, and $D = 1/2^{-1/2}\Gamma(1/2)$. Hence

$$y(0) = B/D = [2^{5/2}\Gamma(1/2)]/[2^{1/2}\Gamma(-3/2)]$$

$$= [2^2\Gamma(1/2)]/[(4/3)\Gamma(1/2)] = 3.$$

We will not include here detailed solutions for Problems 13-21. The solution of Problem 18 is fairly typical.

13. $x^2 J_1(x) + x J_0(x) - \int J_0(x)\,dx + C$

14. $(x^3 - 4x)J_1(x) + 2x^2 J_0(x) + C$

15. $(x^4 - 9x^2)J_1(x) + (3x^3 - 9x)J_0(x) + 9\int J_0(x)\,dx + C$

16. $- x J_0(x) + \int J_0(x)\,dx + C$

17. $2x J_1(x) - x^2 J_0(x) + C$

18. First we integrate by parts with

$$u = x^3 \qquad\qquad dv = J_1(x)$$
$$du = 3x^2 dx \qquad\qquad v = -J_0(x)$$

using Example (1). This yields

$$\int x^3 J_1(x)\,dx = - x^3 J_0(x) + 3 \int x^2 J_0(x)\,dx.$$

Next we integrate by parts with

$$u = x \qquad\qquad dv = x J_0(x)\,dx$$
$$du = dx \qquad\qquad v = x J_1(x)$$

and get

$$\int x^3 J_1(x) \, dx = - x^3 J_0(x) + 3x^2 J_1(x) - 3 \int x \, J_1(x) \, dx.$$

Finally we integrate by parts with

$$u = x \qquad\qquad dv = J_1(x) \, dx$$
$$du = dx \qquad\qquad v = - J_0(x)$$

and get

$$\int x^3 J_1(x) \, dx = - x^3 J_0(x) + 3x^2 J_1(x) + 3x J_0(x) - 3 \int J_0(x) \, dx.$$

19. $(4x^3 - 16x) J_1(x) + (8x^2 - x^4) J_0(x) + C$

20. $- 2J_1(x) + \int J_0(x) \, dx$

21. $J_0(x) - (4/x) J_1(x) + C$

22. Let us define

$$g(x) = \int_0^\pi \cos(x \sin \theta) \, d\theta$$

and note first that

$$g(0) = \int_0^\pi \cos(0) \, d\theta = \pi = \pi \, J_0(0).$$

Differentiation under the integral sign yields

$$g'(x) = - \int_0^\pi \sin(x \sin \theta) \sin \theta \, d\theta.$$

When we integrate by parts with

$$u = \sin(x \sin \theta) \qquad\qquad\qquad dv = \sin \theta \, d\theta$$
$$du = (x \cos \theta) \cos(x \sin \theta) \, d\theta \qquad\qquad v = - \cos \theta$$

we get

$$g'(x) = -x \int_0^\pi \cos^2\theta \, \cos(x \sin \theta) \, d\theta.$$

But differentiation of the first equation for $g'(x)$ yields

$$g''(x) = -\int_0^\pi \sin^2\theta \, \cos(x \sin \theta) \, d\theta.$$

Finally, because $\cos^2\theta + \sin^2\theta = 1$, it follows that

$$g''(x) + (1/x)g'(x) = -\int_0^\pi \cos(x \sin \theta) \, d\theta = -g(x).$$

Thus $y = g(x)$ satisfies Bessel's equation of order zero in the form $y'' + (1/x)y' + y = 0$. Therefore

$$g(x) = a \, J_0(x) + b \, Y_0(x).$$

Since $g(0) = \pi$ is finite and $J_0(0) = 1$, we must have $a = \pi$ and $b = 0$, so $g(x) = \pi \, J_0(x)$, as desired.

23. This is a special case of the discussion below in Problem 24.

24. Given an integer $n \geq 1$, let us define

$$g_n(x) = \int_0^\pi \cos(n\theta - x \sin \theta) \, d\theta.$$

Differentiation yields

$$g_n'(x) = \int_0^\pi \sin(n\theta - x \sin \theta) \sin \theta \, d\theta.$$

Integration by parts with $u = \sin(n\theta - x \sin \theta)$ and $dv = \sin \theta \, d\theta$ yields

$$g_n'(x) = n \int_0^\pi \cos\theta \, \cos(n\theta - x\sin\theta) \, d\theta$$

$$- x \int_0^\pi \cos^2\theta \, \cos(n\theta - x\sin\theta) \, d\theta.$$

But differentiation of the first equation for $g_n'(x)$ yields

$$g_n''(x) = -\int_0^\pi \sin^2\theta \, \cos(n\theta - x\sin\theta) \, d\theta.$$

It follows that

$$g_n''(x) + (1/x)g_n'(x) = -g_n(x)$$

$$+ (n/x) \int_0^\pi \cos\theta \, \cos(n\theta - x\sin\theta) d\theta$$

$$= -g_n(x) - (n/x^2) \int_0^\pi [(n - x\cos\theta) - n]\cos(n\theta - x\sin\theta) d\theta$$

$$= -g_n(x) - (n/x^2) \Big[\sin(n\theta - x\sin\theta)\Big]_0^\pi + (n^2/x^2)g_n(x)$$

$$= -(1 - n^2/x^2)g_n(x).$$

Thus $y = g_n(x)$ satisfies Bessel's equation of order $n \geq 1$. The initial values of $g_n(x)$ are

$$g_n(0) = \int_0^\pi \cos(n\theta) \, d\theta = 0,$$

$$g_n'(0) = \int_0^\pi \sin\theta \, \sin n\theta \, d\theta.$$

If $n = 1$ then $g_1'(0) = \pi/2$, while $g_n'(0) = 0$ if $n \geq 1$. In either case the values of $g_n(0)$ and $g_n'(0)$ are π times those of $J_n(0)$ and $J_n'(0)$, respectively. Now we know from the general solution of Bessel's equation that $g_n(x) = c \, J_n(x)$ for some constant c. If $n = 1$ than the fact that

$$\pi/2 = g_n'(0) = c\, J_1'(0) = c/2$$

implies that $c = \pi$, as desired. But if $n \geq 1$ the fact that

$$0 = g_n'(0) = c\, J_n'(0) = c \cdot 0$$

does not suffice to determine c.

SECTION 3.6

APPLICATIONS OF BESSEL FUNCTIONS

Problems 1-12 are routine applications of the theorem in this section. In each case it is necessary only to identify the coefficients A, B, C and the exponent q in Equation (3); calculate the values of α, β, k, p using the formulas in (5); and then write the general solution specified in Equation (6).

1. $y = x[c_1 J_0(x) + c_2 Y_0(x)]$

2. $y = x^{-1}[c_1 J_1(x) + c_2 Y_1(x)]$

3. $y = x[c_1 J_{1/2}(3x^2) + c_2 J_{-1/2}(3x^2)]$

4. $y = x^3[c_1 J_2(2x^{1/2}) + c_2 Y_2(2x^{1/2})]$

5. $y = x^{-1/3}[c_1 J_{1/3}((1/3)x^{3/2}) + c_2 J_{-1/3}((1/3)x^{3/2})]$

6. $y = x^{-1/4}[c_1 J_0(2x^{3/2}) + c_2 Y_0(2x^{3/2})]$

7. $y = x^{-1}[c_1 J_0(x) + c_2 Y_0(x)]$

8. $y = x^2[c_1 J_1(4x^{1/2}) + c_2 Y_1(4x^{1/2})]$

9. $y = x^{1/2}[c_1 J_{1/2}(2x^{3/2}) + c_2 J_{-1/2}(2x^{3/2})]$

10. $y = x^{-1/4}[c_1 J_{3/2}((2/5)x^{5/2}) + c_2 J_{-3/2}((2/5)x^{5/2})]$

11. $y = x^{1/2}[c_1 J_{1/6}((1/3)x^3) + c_2 J_{-1/6}((1/3)x^3)]$

12. $y = x^{1/2}[c_1 J_{1/5}((4/5)x^{5/2}) + c_2 J_{-1/5}((4/5)x^{5/2})]$

13. We want to solve the equation

$$xy'' + 2y' + xy = 0.$$

If we rewrite it as

$$x^2 y'' + 2xy' + x^2 y = 0$$

we have the form in Equation (3) with $A = 2$, $B = 0$, $C = 1$, and $q = 2$. Then Equation (5) gives $\alpha = -1/2$, $\beta = 1$, $k = 1$, and $p = 1/2$, so by Equation (6) the general solution is

$$y = x^{-1/2}[c_1 J_{1/2}(x) + c_2 J_{-1/2}(x)]$$

$$= x^{-1/2}(2/\pi x)^{1/2}[c_1 \cos x + c_2 \sin x]$$

$$y = x^{-1}(A \cos x + B \sin x),$$

using Equations (19) in Section 3.5.

15. The substitution

$$y = -u'/u, \quad y' = (u')^2/u^2 - u''/u$$

immediately transforms $y' = x^2 + y^2$ to $u'' + x^2u = 0$. The equation

$$x^2u'' + x^4u = 0$$

is of the form in (3) with $A = B = 0$, $C = 1$, and $q = 4$. Equations (5) give $\alpha = 1/2$, $\beta = 2$, $k = 1/2$, and $p = 1/4$, so the general solution is

$$u = x^{1/2}[c_1J_{1/4}(x^2/2) + c_2J_{-1/4}(x^2/2)].$$

To compute u', let $z = x^2/2$ so $x = 2^{1/2}z^{1/2}$. Then Equation (22) in Section 3.5 with $p = 1/4$ yields

$$D_x[x^{1/2}J_{1/4}(x^2/2)] = D_z[2^{1/4}z^{1/4}J_{1/4}(z)] \cdot (dz/dx)$$

$$= 2^{1/4}z^{1/4}J_{-3/4}(x^2/2) \cdot x$$

$$= x^{3/2}J_{-3/4}(x^2/2).$$

Similarly, Equation (23) in Section 3.5 with $p = -1/4$ yields

$$D_x[x^{1/2}J_{-1/4}(x^2/2)] = D_z[2^{1/4}z^{1/4}J_{-1/4}(z)] \cdot (dz/dx)$$

$$= -2^{1/4}z^{1/4}J_{3/4}(z) \cdot x$$

$$= -x^{3/2}J_{3/4}(x^2/2).$$

Therefore

$$u' = x^{3/2}[c_1J_{-3/4}(x^2/2) - c_2J_{3/4}(x^2/2)].$$

It follows finally that the general solution of the Ricatti equation $y' = x^2 + y^2$ is

$$y(x) = -\frac{u'}{u} = x \cdot \frac{J_{3/4}(x^2/2) - c\, J_{-3/4}(x^2/2)}{c\, J_{1/4}(x^2/2) + J_{-1/4}(x^2/2)}$$

where the arbitrary constant is $c = c_1/c_2$.

16. Substitution of the series expressions for the Bessel functions in the formula for $y(x)$ in Problem 15 yields

$$y(x) = x \cdot \frac{A(x^2/2)^{3/4}(1 + \cdots) - cB(x^2/2)^{-3/4}(1 + \cdots)}{cC(x^2/2)^{1/4}(1 + \cdots) + D(x^2/2)^{-1/4}(1 + \cdots)}$$

where each pair of parentheses encloses a power series in x with constant term 1, and

$$A = 1/2^{3/4}\Gamma(7/4) \qquad\qquad B = 2^{3/4}/\Gamma(1/4)$$

$$C = 1/2^{1/4}\Gamma(5/4) \qquad\qquad D = 2^{1/4}/\Gamma(3/4).$$

Multiplication of numerator and denominator by $x^{1/2}$ and a bit of simplification gives

$$y(x) = \frac{2^{-3/4}Ax^3(1 + \cdots) - 2^{3/4}cB(1 + \cdots)}{2^{-1/4}cCx(1 + \cdots) + 2^{1/4}D(1 + \cdots)}.$$

It now follows that

$$y(0) = -2^{1/2}cB/D$$
$$= -2^{1/2}c \cdot [2^{3/4}/\Gamma(1/4)]/[2^{1/4}/\Gamma(3/4)]$$
$$y(0) = -2c \cdot \Gamma(3/4)/\Gamma(1/4). \qquad\qquad (*)$$

(a) If $y(0) = 0$ then $(*)$ gives $c = 0$ in the general solution formula of Problem 15.

(b) If $y(0) = 1$ then $(*)$ gives $c = -\Gamma(1/4)/2\Gamma(3/4)$. More generally, $(*)$ yields the formula

$$y(x) = x \cdot \frac{2\Gamma(3/4)J_{3/4}(x^2/2) + y_0\Gamma(1/4)J_{-3/4}(x^2/2)}{2\Gamma(3/4)J_{-1/4}(x^2/2) - y_0\Gamma(1/4)J_{1/4}(x^2/2)}$$

For the solution of the initial value problem

$$y' = x^2 + y^2, \qquad y(0) = y_0.$$

17. If we write the equation $x^4 y'' + \gamma^2 y = 0$ in the form

$$x^2 y'' + \gamma^2 x^{-2} y = 0$$

we see that it is of the form in Equation (3) of this section with $A = B = 0$, $C = \gamma^2$, and $q = -2$. Then Equations (5) give $\alpha = 1/2$, $\beta = -1$, $k = \gamma$, and $p = -1/2$, so the theorem yields the general solution

$$y(x) = x^{1/2}[c_1 J_{1/2}(\gamma/x) + c_2 J_{-1/2}(\gamma/x)]$$

$$= x[A\cos(\gamma/x) + B\sin(\gamma/x)],$$

using Equations (19) in Section 3.5 for $J_{1/2}(x)$ and $J_{-1/2}(x)$. With a and b both nonzero, the initial conditions $y(a) = y(b) = 0$ yield the equations

$$A\cos(\gamma/a) + B\sin(\gamma/a) = 0$$
$$A\cos(\gamma/b) + B\sin(\gamma/b) = 0.$$

These equations have a nontrivial solution for A and B only if the coefficient determinant

$$\Delta = \sin(\gamma/b)\cos(\gamma/a) - \sin(\gamma/a)\cos(\gamma/b)$$
$$= \sin(\gamma/b - \gamma/a) = \sin(\gamma L/ab)$$

is nonzero. Hence $\gamma L/ab$ must be an integral multiple $n\pi$ of π, and then the nth buckling force is

$$P_n = EI_0 \gamma_n^2 / b^4$$

$$= EI_0(n\pi ab/L)^2 b^4 = (n^2\pi^2 EI_0/L^2)(a/b)^2.$$

18. The substitution $L = a + bt$ in $L\theta'' + 2L'\theta' + g\theta = 0$ yields the transformed equation

$$L^2\theta''(L) + 2L\theta'(L) + (g/b^2)L\theta = 0$$

with independent variable L that is of the form in (3) with $A = 2$, $B = 0$, $q = 1$, and $C = g/b^2$. Hence

$$\alpha = -1/2, \quad \beta = 1/2, \quad k = 2g^{1/2}/b, \quad \text{and} \quad p = 1,$$

so

$$\theta(L) = L^{-1/2}[AJ_1(2g^{1/2}L^{1/2}/b) + BY_1(2g^{1/2}L^{1/2}/b)].$$

SECTION 3.7

APPENDIX ON INFINITE SERIES AND THE ATOM

We offer this section as a dramatic illustration of the fact that "differential equations explain the universe." Although in typical courses class time may not be available for "infinite series and the atom," we think it beneficial for students to see in the text supplementary reading material on the role of series solutions in one of the great triumphs of modern science. It should be emphasized that no prior knowledge of quantum theory is assumed. Indeed, we think the approach in this section is (for mathematics students) the best introduction to the subject.

CHAPTER 4

THE LAPLACE TRANSFORM

SECTION 4.1

LAPLACE TRANSFORMS AND INVERSE TRANSFORMS

The objectives of this section are especially clearcut. They
include familiarity with the definition of the Laplace transform
$\mathcal{L}\{f(t)\} = F(s)$ that is given in Equation (1); the direct
application of this definition to calculate Laplace transforms of
simple functions (as in Examples 1-3); and the use of known
transforms (those listed in Figure 4.2) to find Laplace
transforms and inverse transforms (as in Examples 4-6). Perhaps
students need to be told explicitly to memorize the transforms
that are listed in the short table that appears in Figure 4.2.

1. $\quad \mathcal{L}\{t\} = \displaystyle\int_0^\infty e^{-st} t \; dt \qquad\qquad (u = -st, \quad du = -s \; dt)$

$\qquad = \displaystyle\int_0^{-\infty} \left[\frac{1}{s^2}\right] u e^u \; du = \frac{1}{s^2} \left[(u - 1)e^u\right]_0^{-\infty} = \frac{1}{s^2}$

2. $\quad$ We substitute $\; u = -st \;$ in the tabulated integral

$$\int u^2 e^u du = e^u(u^2 - 2u + 2) + c$$

$\quad$ (or, alternatively, integrate by parts) and get

$$\mathscr{L}\{t^2\} = \int_0^\infty e^{-st}t^2 dt = \left[-e^{-st}(t^2/s + 2t/s^2 + 2/s^3) \right]_{t=0}^\infty = \frac{2}{s^3}$$

3. $\mathscr{L}\{e^{3t+1}\} = \int_0^\infty e^{-st}e^{3t+1}dt$

$$= e \int_0^\infty e^{-(s-3)t}dt = \frac{e}{s-3}$$

4. With $a = -s$ and $b = 1$ the tabulated integral

$$\int e^{au}\cos bu \; du = e^{au}\left[\frac{a \cos bu + b \sin bu}{(a^2 + b^2)}\right] + C$$

yields

$$\mathscr{L}\{\cos t\} = \int_0^\infty e^{-st}\cos t \; dt$$

$$= \left[e^{-st}(-s\cos t + \sin t)/(s^2 + 1)\right]_{t=0}^\infty = \frac{s}{s^2 + 1}.$$

5. $\mathscr{L}\{\sinh t\} = (1/2)\mathscr{L}\{e^t - e^{-t}\}$

$$= \frac{1}{2}\int_0^\infty e^{-st}(e^t - e^{-t})dt$$

$$= \frac{1}{2}\int_0^\infty \left[e^{-(s-1)t} - e^{-(s+1)t}\right]dt$$

$$= (1/2)[1/(s-1) - 1/(s+1)] = 1/(s^2 - 1)$$

6. $\mathscr{L}\{\sin^2 t\} = \int_0^\infty e^{-st}\sin^2 t \; dt = \frac{1}{2}\int_0^\infty e^{-st}(1 - \cos 2t)dt$

$$= \frac{1}{2}\left[e^{-st}(-\frac{1}{s}) - e^{-st} \cdot \frac{-s\cos 2t + 2\sin 2t}{s^2 + 4}\right]_{t=0}^\infty$$

$$\mathscr{L}\{\sin^2 t\} = \frac{1}{2}\left[\frac{1}{s} - \frac{s}{s^2 + 4}\right]$$

7. $\mathcal{L}\{f(t)\} = \int_0^1 e^{-st}dt = [-\frac{1}{s}e^{-st}]_0^1 = (1 - e^{-s})/s$

8. $\mathcal{L}\{f(t)\} = \int_1^2 e^{-st}dt = [-e^{-st}/s]_1^2 = (e^{-s} - e^{-2s})/s$

9. $\mathcal{L}\{f(t)\} = \int_0^1 e^{-st}t \ dt = (1 - e^{-s} - se^{-s})/s^2$

10. $\mathcal{L}\{f(t)\} = \int_0^1 (1 - t)e^{-st}dt = [-e^{-st}(1/s + t/s + 1/s^2)]_0^1$

    $= (1/s + 1/s^2) - e^{-s}(2/s + 1/s^2)$

11. $s^{-3/2}\Gamma(3/2) + 3s^{-2} = (1/2)\sqrt{\pi}s^{-3/2} + 3s^{-2}$

12. $3 \cdot \Gamma(7/2)/s^{7/2} - 4 \cdot 3!/s^4 = 45\sqrt{\pi}/8s^{7/2} - 24/s^4$

13. $\mathcal{L}\{t - 2e^{3t}\} = 1/s^2 - 2/(s - 3)$

14. $\mathcal{L}\{t^{3/2} + e^{-10t}\} = 3\sqrt{\pi}/4s^{5/2} + 1/(s + 10)$

15. $\mathcal{L}\{1 + \cosh 5t\} = 1/s + s/(s^2 - 25)$

16. $\mathcal{L}\{\sin 2t + \cos 2t\} = (s + 2)/(s^2 + 4)$

17. $\mathcal{L}\{\cos^2 2t\} = (1/2)\mathcal{L}\{1 + \cos 4t\} = (1/2)[1/s + s/(s^2 + 16)]$

18. $\mathcal{L}\{\sin 3t \cos 3t\} = \mathcal{L}\{(1/2)\sin 6t\} = 3/(s^2 + 36)$

19. $\mathscr{L}\{(1 + t)^3\} = \mathscr{L}\{1 + 3t + 3t^2 + t^3\}$

$$= 1/s + 3/s^2 + 6/s^3 + 6/s^4$$

20. Integrating by parts with $u = t$, $dv = e^{-(s-1)t}dt$, we get

$$\mathscr{L}\{te^t\} = \int_0^\infty te^t\, e^{-st}dt = \int_0^\infty te^{-(s-1)t}dt$$

$$= \left[\, - te^{-(s-1)t}/(s - 1)\right]_0^\infty + \frac{1}{s-1}\int_0^\infty e^t e^{-st}dt$$

$$= [1/(s - 1)]\, \mathscr{L}\{e^t\} = 1/(s - 1)^2$$

21. Integration by parts with $u = t$ and $dv = e^{st}\cos 2t\, dt$ yields

$$\mathscr{L}\{t\,\cos\,2t\} = \int_0^\infty te^{-st}\cos\,2t\,dt$$

$$= -[1/(s^2 + 4)]\int_0^\infty e^{-st}(-s\,\cos\,2t + 2\,\sin\,2t)dt$$

$$= -[1/(s^2 + 4)][(-s)\mathscr{L}\{\cos\,2t\} + 2\mathscr{L}\{\sin\,2t\}]$$

$$= -[1/(s^2 + 4)][(-s^2)/(s^2 + 4) + 4/(s^2 + 4)]$$

$$= (s^2 - 4)/(s^2 + 4)^2$$

22. $\mathscr{L}\{\sinh^2 3t\} = \mathscr{L}\{(1/2)(\cosh\,6t - 1)\}$

$$= (1/2)[s/(s^2 - 36) - 1/s]$$

23. $\mathscr{L}^{-1}\{3/s^4\} = t^3/2$

24. $\mathscr{L}^{-1}\{s^{-3/2}\} = 2t^{1/2}/\sqrt{\pi}$

25. $\mathscr{L}^{-1}\{1/s - 2/s^{5/2}\} = 1 - 8t^{3/2}/3\sqrt{\pi}$

26. $\mathscr{L}^{-1}\{1/(s + 5)\} = e^{-5t}$

27. $\mathscr{L}^{-1}\{3/(s - 4)\} = 3e^{4t}$

28. $\mathscr{L}^{-1}\{(3s + 1)/(s^2 + 4)\}$

$$= 3\mathscr{L}^{-1}\{s/(s^2 + 4)\} + (1/2)\mathscr{L}^{-1}\{2/(s^2 + 4)\}$$

$$= 3 \cos 2t + (1/2)\sin 2t$$

29. $\mathscr{L}^{-1}\{5 - 3s)/(s^2 + 9)\} = (5/3)\sin 3t - 3 \cos 3t$

30. $\mathscr{L}^{-1}\{(9 + s)/(4 - s^2)\}$

$$= -\mathscr{L}^{-1}\{s/(s^2 - 4)\} - (9/2)\mathscr{L}^{-1}\{2/(s^2 - 4)\}$$

$$= - \cosh 2t - (9/2)\sinh 2t$$

31. $\mathscr{L}^{-1}\{(10s - 3)/(25 - s^2)\} = (3/5)\sinh 5t - 10 \cosh 5t$

32. $\mathscr{L}^{-1}\{2s^{-1}e^{-3s}\} = 2\ u(t - 3)$

34. $\mathscr{L}\{\sinh kt\} = (1/2)\mathscr{L}\{e^{kt}\} - (1/2)\mathscr{L}\{e^{-kt}\}$

$$= (1/2)[1/(s - k) - 1/(s + k)] = k/(s^2 - k^2)$$

37. $f(t) = 1 - u_a(t) = 1 - u(t - a)$

38. $\mathscr{L}\{f(t)\} = \mathscr{L}\{u(t - a) - u(t - b)\} = s^{-1}(e^{-as} - e^{-bs})$

39. $\mathscr{L}\{f(t)\} = \sum \mathscr{L}\{u(t - n)\}$

$$= \sum s^{-1}e^{-ns}$$

$$= s^{-1}(1 + e^{-s} + e^{-2s} + e^{-3s} + \cdots)$$

$$\mathcal{L}\{f(t)\} = s^{-1}/(1 - e^{-s})$$

40. $\mathcal{L}\{f(t)\} = \sum (-1)^n \mathcal{L}\{u(t - n)\}$

$\qquad = \sum (-1)^n s^{-1} e^{-ns}$

$\qquad = s^{-1}(1 - e^{-s} + e^{-2s} - e^{-3s} + \cdots)$

$\mathcal{L}\{f(t)\} = s^{-1}/(1 + e^{-s})$

41. $\mathcal{L}\{g(t)\} = \mathcal{L}\{2f(t) - 1\}$

$\qquad = 2/s(1 + e^{-s}) - 1/s$

$\qquad = s^{-1}[(1 - e^{-s})/(1 + e^{-s})]$

$\qquad = s^{-1}[(e^{s/2} - e^{-s/2})/(e^{s/2} + e^{-s/2})]$

$\mathcal{L}\{g(t)\} = s^{-1}\tanh(s/2)$

SECTION 4.2

TRANSFORMATION OF INITIAL VALUE PROBLEMS

The focus of this section is on the use of transforms of
derivatives (Theorem 1) to solve initial value problems (as in
Examples 1 and 2). Transforms of integrals (Theorem 2) appear
less frequently in practice, and the extension of Theorem 1 at
the end of Section 4.2 may be considered entirely optional
(except perhaps for electrical engineering students).

1. $[s^2 X(s) - 5s] + 4\{X(s)\} = 0$

$\qquad X(s) = 5s/(s^2 + 4) = 5[s/(s^2 + 4)]$

$\qquad x(t) = \mathcal{L}^{-1}\{X(s)\} = 5 \cos 2t$

2. $[s^2 X(s) - 3s - 4] + 9[X(s)] = 0$

$\qquad X(s) = (3s + 4)/(s^2 + 9) = 3[s/(s^2 + 9)] + (4/3)[3/(s^2 + 9)]$

$x(t) = \mathcal{L}^{-1}\{X(s)\} = 3 \cos 3t + (4/3)\sin 3t$

3. $[s^2X(s) - 2] - [sX(s)] - 2[X(s)] = 0$

$X(s) = 2/(s^2 - s - 2) = (2/3)[1/(s - 2) - 1/(s + 1)]$

$x(t) = (2/3)(e^{2t} - e^{-t})$

4. $[s^2X(s) - 2s + 3] + 8[s\ X(s) - 2] + 15[X(s)] = 0$

$X(s) = (2s + 13)/(s^2 + 8s + 15)$

$= (7/2)/(s + 3) - (3/2)/(s + 5)$

$x(t) = \mathcal{L}^{-1}\{X(s)\} = (7/2)e^{-3t} - (3/2)e^{-5t}$

5. $[s^2X(s)] + [X(s)] = 2/(s^2 + 4)$

$X(s) = 2/[(s^2 + 1)(s^2 + 4)]$

$= (1/3)[2/(s^2 + 1) - 2/(s^2 + 4)]$

$x(t) = (2 \sin t - \sin 2t)/3$

6. $[s^2X(s)] + 4[X(s)] = \mathcal{L}\{\cos t\} = s/(s^2 + 1)$

$X(s) = s/(s^2 + 1)(s^2 + 4) = (1/3)s/(s^2 + 1) - (1/3)s/(s^2 + 4)$

$x(t) = \mathcal{L}^{-1}\{X(s)\} = (\cos t - \cos 2t)/3$

7. $[s^2X(s) - s] + [X(s)] = s/s^2 + 9)$

$(s^2 + 1)X(s) = s + s/(s^2 + 9) = (s^3 + 10s)/(s^2 + 9)$

$X(s) = (s^3 + 10s)/[(s^2 + 1)(s^2 + 9)]$

$= (1/8)[9s/(s^2 + 1) - s/(s^2 + 9)]$

$x(t) = (9 \cos t - \cos 3t)/8$

8. $[s^2X(s)] + 9[X(s)] = \mathcal{L}\{1\} = 1/s$

$X(s) = 1/s(s^2 + 9) = (1/9)/s - (1/9)s/(s^2 + 9)$

$$x(t) = \mathscr{L}^{-1}\{X(s)\} = (1 - \cos 3t)/9$$

9. $s^2X(s) + 4sX(s) + 3X(s) = 1/s$

    $X(s) = 1/s(s^2 + 4s + 3)$

    $\qquad = (1/6)[2/s - 3/(s + 1) + 1/(s + 3)]$

    $x(t) = (2 - 3e^{-t} + e^{-3t})/6$

10. $[s^2X(s) - 2] + 3[sX(s)] + 2[X(s)] = \mathscr{L}\{t\} = 1/s^2$

    $(s^2 + 3s + 2)X(s) = 2 + 1/s^2 = (2s^2 + 1)/s^2$

    $X(s) = (2s^2 + 1)/s^2(s^2 + 3s + 2)$

    $\qquad = - (3/4)/s + (1/2)/s^2 + 3/(s + 1) - (9/4)/(s + 2)$

    $x(t) = \mathscr{L}^{-1}\{X(s)\} = (-3 + 2t + 12e^{-t} - 9e^{-2t})/4$

11. $f(t) = \displaystyle\int_0^t e^{3t}dt = (e^{3t} - 1)/3$

12. $f(t) = \displaystyle\int_0^t 3e^{-5t}dt = 3(1 - e^{-5t})/5$

13. $f(t) = \displaystyle\int_0^t (1/2)\sin 2t \; dt = (1 - \cos 2t)/4$

14. $f(t) = \displaystyle\int_0^t [2 \cos 3t + (1/3)\sin 3t]dt$

    $\qquad = (6 \sin 3t - \cos 3t + 1)/9$

15. $f(t) = \displaystyle\int_0^t \left[\int_0^t \sin t \; dt \right] dt$

    $\qquad = \displaystyle\int_0^t (1 - \cos t)dt = t - \sin t$

16. $f(t) = \int_0^t (1/3) \sinh 3t \, dt = (\cosh 3t - 1)/9$

17. $f(t) = \int_0^t [\int_0^t \sinh t \, dt] \, dt$

$= \int_0^t (\cosh t - 1) dt = \sinh t - t$

18. $f(t) = \int_0^t (e^{-t} - e^{-2t}) dt = (e^{-2t} - 2e^{-t} + 1)/2$

19. With $f(t) = \cos kt$ and $F(s) = s/(s^2 + k^2)$, Theorem 1 yields

$$\mathcal{L}\{-k \sin kt\} = \mathcal{L}\{f'(t)\} = sF(s) - 1$$
$$= [s^2/(s^2 + k^2)] - 1$$
$$= -k^2/(s^2 + k^2)$$

so division by $-k$ yields $\mathcal{L}\{\sin kt\} = k/(s^2 + k^2)$.

20. With $f(t) = \sinh kt$ and $F(s) = k/(s^2 - k^2)$, Theorem 1 yields

$$\mathcal{L}\{f'(t)\} = \mathcal{L}\{k \cosh kt\} = ks/(s^2 - k^2) = sF(s),$$

so it follows that $\mathcal{L}\{\cosh kt\} = s/(s^2 - k^2)$.

21. (a) With $f(t) = t^n e^{at}$ and $f'(t) = nt^{n-1}e^{at} + at^n e^{at}$, Theorem 1 yields

$$\mathcal{L}\{nt^{n-1}e^{at} + at^n e^{at}\} = s\mathcal{L}\{t^n e^{at}\}$$

so

$$n\mathcal{L}\{t^{n-1}e^{at}\} = (s - a)\mathcal{L}\{t^n e^{at}\}$$

24. Problems 22 and 24 are the trigonometric and hyperbolic

versions of essentially the same computation. For Problem 24 we let $f(t) = t \cosh kt$, so $f(0) = 0$. Then

$$f'(t) = \cosh kt + kt \sinh kt,$$
$$f''(t) = 2k \sinh kt + k^2 t \cosh kt,$$

and $f'(0) = 1$, so Formula (5) yields

$$\mathcal{L}\{2k \sinh kt + k^2 t \cosh kt\} = s^2 \mathcal{L}\{t \cosh kt\} - 1,$$

$$2k^2/(s^2 - k^2) = k^2 F(s) = s^2 F(s) - 1.$$

We readily solve this last equation for

$$\mathcal{L}\{t \cosh kt\} = F(s) = (s^2 + k^2)/(s^2 - k^2)^2.$$

26. If $f(t) = u(t - a)$, then the only jump in $f(t)$ is $j_1 = 1$ at $t_1 = a$. Since $f(0) = 0$ and $f'(t) = 0$, Formula (18) yields

$$0 = s\, F(s) - 0 - e^{as}(1).$$

Hence $\mathcal{L}\{u(t - a)\} = F(s) = s^{-1} e^{-as}$.

28. The square wave function of Figure 4.14 has a sequence $\{t_n\}$ of jumps with $t_n = n$ and $j_n = 2(-1)^n$ for $n = 1, 2, 3, \cdots$. Hence Formula (18) yields

$$0 = s\, F(s) - 1 - \sum (-1)^n 2 e^{-ns}.$$

It follows that

$$
\begin{aligned}
s\, F(s) &= 1 + 2 \sum (-1)^n e^{-ns} \\
&= -1 + 2(1 - e^{-s} + e^{-2s} - e^{-3s} + \cdots) \\
&= -1 + 2/(1 + e^{-s}) \\
&= (1 - e^{-s})/(1 + e^{-s})
\end{aligned}
$$

$$= (e^{s/2} - e^{-s/2}) / (e^{s/2} + e^{-s/2})$$
$$s \, F(s) = \tanh (s/2),$$

because $2 \cosh (s/2) = e^{s/2} + e^{-s/2}$ and $2 \sinh (s/2) = e^{s/2} - e^{-s/2}$.

30. If $g(t)$ is the triangular wave function of Figure 4.16 and $f(t)$ is the square wave function of Problem 28, then $g'(t) = f(t)$. Hence Theorem 1 and the result of Problem 28 yield

$$\mathcal{L}\{g'(t)\} = s \, \mathcal{L}\{g(t)\} - g(0),$$
$$F(s) = s \, G(s),$$
$$\mathcal{L}\{g(t)\} = s^{-1}F(s) = s^{-2}\tanh(s/2).$$

SECTION 4.3

TRANSLATION AND PARTIAL FRACTIONS

This section is devoted to the computational nuts and bolts of the staple technique for the inversion of Laplace transforms -- partial fraction decompositions. If time does not permit going further in this chapter, Sections 4.1-4.3 provide a self-contained introduction to Laplace transforms that suffices for the most common elementary applications.

1. $$\mathcal{L}\{t^4\} = 24/s^5,$$

so $$\mathcal{L}\{t^4 e^{\pi t}\} = 24/(s - \pi)^5.$$

2. $$\mathcal{L}\{t^{3/2}\} = (3\sqrt{\pi}/4) s^{-5/2},$$

so $$\mathcal{L}\{t^{3/2}e^{-4t}\} = (3\sqrt{\pi}/4)(s + 4)^{-5/2}.$$

3. $$\mathcal{L}\{\sin 3\pi t\} = 3\pi/(s^2 + 9\pi^2),$$

so $$\mathcal{L}\{e^{-2t}\sin 3\pi t\} = 3\pi/[(s + 2)^2 + 9\pi^2].$$

4. $\cos(2t - \pi/4) = (1/\sqrt{2})(\cos 2t + \sin 2t)$

$\mathcal{L}\{\cos(2t - \pi/4)\} = (1/\sqrt{2})(s + 2)/(s^2 + 4)$

$\mathcal{L}\{f(t)\} = (1/\sqrt{2})(s + 1/2 + 2)/[(s + 1/2)^2 + 4]$

$= (\sqrt{2})(2s + 5)/(4s^2 + 4s + 17)$

5. $F(s) = 3/(2s - 4) = (3/2)/(s - 2)$

$f(t) = (3/2)e^{2t}$

6. $F(s) = [(s + 1) - 2]/(s + 1)^3 = 1/(s + 1)^2 - 2/(s + 1)^3$

$f(t) = e^{-t}(t - t^2)$

7. $F(s) = 1/(s + 2)^2$, $\quad f(t) = te^{-2t}$

8. $F(s) = (s + 2)/[(s + 2)^2 + 1]$; $\quad f(t) = e^{-2t}\cos t$

9. $F(s) = [3(s - 3) + (7/2)(4)]/[(s - 3)^2 + 16]$

$f(t) = e^{3t}[3 \cos 4t + (7/2)\sin 4t]$

10. $F(s) = (2s - 3)/[(3s - 2)^2 + 16]$

$= [(2/9)(s - 2/3) - (5/36)(4/3)]/[(s - 2/3)^2 + (4/3)^2]$

$f(t) = (1/36)e^{2t/3}(8 \cos 4t/3 - 5 \sin 4t/3)$

11. $f(t) = \mathcal{L}^{-1}\{(1/4)[1/(s - 2) - 1/(s + 2)]\}$

$= (1/4)(e^{2t} - e^{-2t}) = (1/2)\sinh 2t$

12. $f(t) = \mathcal{L}^{-1}\{2/s + 3/(s - 3)\} = 2 + 3e^{3t}$

13. $f(t) = \mathcal{L}^{-1}\{3/(s + 2) - 5/(s + 5)\} = 3e^{-2t} - 5e^{-5t}$

14. $F(s) = 2/s + 1/(s - 2) - 3/(s + 1)$

$f(t) = 2 + e^{2t} - 3e^{-t}$

15. $f(t) = \mathcal{L}^{-1}\{(1/25)[1/(s - 5) - 1/s - 5/s^2]\}$

$= (e^{5t} - 1 - 5t)/25$

16. $F(s) = (1/125)[- 2/(s - 2) + 5/(s - 2)^2$

$+ 2/(s + 3) + 5/(s + 3)^2]$

$f(t) = (1/125)[e^{2t}(- 2 + 5t) + e^{-3t}(2 + 5t)]$

17. $f(t) = (1/8)\mathcal{L}^{-1}\{1/(s^2 - 4) - 1/(s^2 + 4)\}$

$= (\sinh 2t - \sin 2t)/16$

18. $F(s) = 6/(s - 4) + 12/(s - 4)^2 + 48/(s - 4)^3 + 64/(s - 4)^4$

$f(t) = e^{4t}(6 + 12t + 24t^2 + 32t^3/3)$

19. $F(s) = (1/3)[(2s + 4)/(s^2 + 4) - (2s + 1)/s^2 + 1)]$

$f(t) = (1/3)(2 \cos 2t + 2 \sin 2t - 2 \cos t - \sin t)$

20. $F(s) = (1/32)[- 1/(s - 2) + 2/(s - 2)^2$

$+ 1/(s + 2) + 2/(s + 2)^2]$

$f(t) = (1/32)[e^{2t}(- 1 + 2t) + e^{-2t}(1 + 2t)]$

21. $f(t) = (1/2)e^{-t}(5 \sin t - 3t \cos t - 2t \sin t)$

22. First we need to find A, B, C, D so that

$$\frac{2s^3 - s^2}{(4s^2 - 4s + 5)^2} = \frac{As + B}{4s^2 - 4s + 5} + \frac{Cs + D}{(4s^2 - 4s + 5)^2} \ .$$

When we multiply each side by the quadratic factor we get the identity

$$2s^3 - s^2 = (As + B)(4s^2 - 4s + 5) + Cs + D.$$

When we substitute the root $s = 1/2 + i$ of the quadratic into this identity, we find that $C = -3/2$ and $D = -5/4$. When we first differentiate each side of the identity and then substitute the root, we find that $A = 1/2$ and $B = 1/4$. Writing

$$4s^2 - 4s + 5 = 4[(s - 1/2)^2 + 1],$$

it follows that

$$F(s) = \frac{1}{8} \cdot \frac{(s - 1/2) + 1}{(s - 1/2)^2 + 1} - \frac{1}{32} \cdot \frac{3(s - 1/2) + 4}{[(s - 1/2)^2 + 1]^2}.$$

Finally the results

$$\mathcal{L}^{-1}\{2s/(s^2 + 1)^2\} = t \sin t,$$
$$\mathcal{L}^{-1}\{2/(s^2 + 1)^2\} = \sin t - t \cos t$$

of Example 4 and Problem 25 in Section 4.2, together with the translation theorem, yield

$$f(t) = (1/64)e^{t/2}[(8 + 4t)\cos t + (4 - 3t)\sin t].$$

24.
$$\frac{4as}{s^4 + 4a^4} = \frac{1}{s^2 - 2as + 2a^2} - \frac{1}{s^2 + 2as + 2a^2}$$

and $s^2 \pm 2as + 2a^2 = (s \pm a)^2 + a^2$, so it follows that

$$\mathcal{L}^{-1}\{s/(s^4 + 4a^4)\} = (1/4a^2)(e^{at} - e^{-at})\sin at$$

$$= (1/2a^2)\sinh at \sin at$$

27. $[s^2X(s) - 2s - 3] + 6[sX(s) - 2] + 2sX(s) = 0$

$$X(s) = [2(s + 3) + 9]/[(s + 3)^2 + 16]$$

$$x(t) = e^{-3t}[2 \cos 4t + (9/4)\sin 4t]$$

28. $X(s) = 2/s(s - 2)(s - 4)$

    $$= (1/4)/s - (1/2)/(s - 2) + (1/4)/(s - 4)$$

    $$x(t) = (1 - 2e^{2t} + e^{4t})/4$$

29. $(s^2 - 4)X(s) = 3/s^2$

    $$X(s) = (3/4)[1/(s^2 - 4) - 1/s^2]$$

    $$x(t) = 3 \sinh 2t - 6t)/8$$

30. $X(s) = 1/(s + 1)(s^2 + 4s + 8)$

    $$= (1/5)/(s + 1) - (1/5)(s + 3)/[(s + 2)^2 + 4]$$

    $$x(t) = (1/10)[2e^{-t} - e^{-2t}(2 \cos 2t + \sin 2t)]$$

31. $[s^3X(s) - s - 1] + [s^2X(s) - 1] - 6[sX(s)] = 0$

    $$X(s) = (1/15)[-5/s + 6/(s - 2) - 1/(s + 3)]$$

    $$x(t) = (-5 + 6e^{2t} - e^{-3t})/15$$

32. $X(s) = s^3/(s^4 - 1) = (1/2)s/(s^2 + 1) + (1/2)s/(s^2 - 1)$

    $$x(t) = (\cos t + \cosh t)/2$$

33. $[s^4X(s) - 1] + X(s) = 0$

    $$X(s) = 1/(s^4 + 1)$$

    It therefore follows from Problem 26 with $a = 1/\sqrt{2}$ that

    $$x(t) = a(\cosh at \sin at - \sinh at \cos at).$$

34. $\{s^4X(s) - 2s^2 + 13] + 13[s^2X(s) - 2] + 36 \ X(s) = 0$

Section 4.3

157

$X(s) = (2s^2 + 13)/(s^4 + 13s^2 + 36) = 1/(s^2 + 4) + 1/(s^2 + 9)$

$x(t) = (1/2)\sin 2t + (1/3)\sin 3t$

35. $X(s) = 1/(s^2 + 4)^2$, so Equation (16) yields

   $x(t) = (\sin 2t - 2t \cos 2t)/16$

36. $X(s) = 1/(s - 2)(s^4 + 2s^2 + 1)$

$$= \frac{1}{25} \cdot \frac{1}{s - 2} - \frac{1}{25} \cdot \frac{s + 2}{s^2 + 1} - \frac{1}{5} \cdot \frac{s + 2}{(s^2 + 1)^2}$$

$x(t) = (1/25)e^{2t} - (1/25)\cos t - (2/25)\sin t$

$\qquad - (1/10)(t \sin t) - (1/5)(\sin t - t \cos t)$

$\qquad = (1/50)[2e^{2t} + (10t - 2)\cos t - (5t + 14)\sin t]$

37. $X(s) = (2s^2 + 4s + 3)/[(s + 1)^2(s^2 + 4s + 13)]$

   $= [-1/(s + 1) + 5/(s + 1)^2 + (s + 98)/(s^2 + 4s + 13)]/50$

   $x(t) = [(-1 + 5t)e^{-t} + e^{-2t}(\cos 3t + 32 \sin 3t)]/50$

38. $[s^2X(0) - s + 1] + 6[s\,X(s) - 1] + 18X(s) = s/(s^2 + 4)$

   $X(s) = (s + 5)/(s^2 + 6s + 18)$

   $\qquad\qquad\qquad + s/[(s^2 + 4)(s^2 + 6s + 18)]$

$$= \frac{1}{170} \cdot \frac{163(s + 3) + 307}{(s + 3)^2 + 9} + \frac{1}{170} \cdot \frac{7s + 12}{s^2 + 4}$$

   $x(t) = (1/510)e^{-3t}(489 \cos 3t + 307 \sin 3t)$

   $\qquad\qquad + (1/170)(7 \cos 2t + 6 \sin 2t)$

39. $x'' + 9x = 6 \cos 3t$

   $(s^2 + 9)X(s) = 6s/(s^2 + 9)$

   $x(t) = \mathcal{L}^{-1}\{6s/(s^2 + 9)^2\} = t \sin 3t$

40. $x'' + 0.4x' + 9.04x = 6e^{-t/5}\cos 3t$

$$(s^2 + 0.4s + 9.04)X(s) = 6(s + 0.2)/[s + 0.2)^2 + 9]$$
$$x(t) = \mathscr{L}^{-1}\{6(s + 0.2)/[(s + 0.2)^2 + 9]^2\}$$
$$= te^{-t/5}\sin 3t$$

SECTION 4.4

DERIVATIVES, INTEGRALS, AND PRODUCTS OF TRANSFORMS

This section completes the presentation of the standard "operational properties" of Laplace transforms, the most important one here being the convolution property $\mathscr{L}\{f*g\} = \mathscr{L}\{f\}\cdot\mathscr{L}\{g\}$, where the convolution $f*g$ is defined by

$$f*g(t) = \int_0^t f(x)g(t - x)\,dx.$$

In an applied course the proofs may (as marked) be regarded as optional.

1. $t*1 = t^2/2$

2. To compute $\int xe^{a(t-x)}dx = e^{at}\int xe^{-ax}dx$, substitute $u = -ax$ and use the integral formula

    $$\int ue^u du = (u - 1)e^u + C.$$

    Answer: $(e^{at} - at - 1)/a^2$

3. To compute $\int \sin x \sin(t - x)\,dx$, first apply the identity

    $$\sin A \sin B = [\cos(A - B) - \cos(A + B)]/2.$$

    Answer: $(\sin t - t \cos t)/2$

4. To compute $\int x^2\cos(t - x)\,dx$, first substitute

$$\cos(t - x) = \cos t \cos x + \sin t \sin x,$$

and then use the integral formulas

$$\int x^2 \cos x \, dx = \quad x^2 \sin x + 2x \cos x - 2 \sin x + C,$$

$$\int x^2 \sin x \, dx = -x^2 \cos x + 2x \sin x + 2 \cos x + C.$$

Answer: $2(t - \sin t)$

5. $e^{at} * e^{at} = t e^{at}$

6. $(e^{at} - e^{bt}) / (a - b)$

7. $f(t) = 1 * e^{3t} = (e^{3t} - 1)/3$

8. $f(t) = (1) * (\sin 2t)/2 = (1 - \cos 2t)/4$

9. $f(t) = (1/9)(\sin 3t) * (\sin 3t) = (\sin 3t - 3t \cos 3t)/54$

10. $f(t) = (t) * (\sin kt)/k = (kt - \sin kt)/k^3$

11. $f(t) = (\cos 2t) * (\cos 2t) = (\sin 2t + 2t \cos 2t)/4$

12. $f(t) = (e^{-2t} \sin t) * (1)$

$$= \int_0^t e^{-2x} \sin x \, dx$$

$$= (1/5)[1 - e^{-2t}(\cos t + 2 \sin t)]$$

13. $f(t) = e^{3t} * \cos t$
    $$= (-3 \cos t + \sin t + 3e^{3t})/10$$

14. $f(t) = (\cos 2t) * (\sin t) = (\cos t - \cos 2t)/3$ with much use of the trigonometric addition formulas.

15. $\mathcal{L}\{t \sin t\} = -(d/ds)\mathcal{L}\{\sin 3t\}$

$= -(d/ds)(3/(s^2 + 9))$

$= 6s/(s^2 + 9)^2$

16. $\mathcal{L}\{t^2\cos 2t\} = (d^2/ds^2)\mathcal{L}\{\cos 2t\}$

$= (d^2/ds^2)[s/(s^2 + 4)]$

$= (2s^3 - 24s)/(s^2 + 4)^3$

17. $\mathcal{L}\{e^{2t}\cos 3t\} = (s - 2)/(s^2 - 4s + 13)$

$\mathcal{L}\{te^{2t}\cos 3t\} = -(d/ds)[(s - 2)/(s^2 - 4s + 13)]$

$= (s^2 - 4s - 5)/(s^2 - 4s + 13)^2$

18. $\mathcal{L}\{\sin^2 t\} = \mathcal{L}\{(1 - \cos 2t)/2\} = 2/s(s^2 + 4)$

$\mathcal{L}\{e^{-t}\sin^2 t\} = 2/(s + 1)(s^2 + 2s + 5)$

$\mathcal{L}\{te^{-t}\sin^2 t\} = -(d/ds)[2/(s + 1)(s^2 + 2s + 5)]$

$= 2(3s^2 + 6s + 7)/(s + 1)^2(s^2 + 2s + 5)^2$

19. $\mathcal{L}\{(\sin t)/t\} = \int_s^\infty [1/(s^2 + 1)]ds$

$= [\tan^{-1}s]_s^\infty$

$= (\pi/2) - \tan^{-1}s = \tan^{-1}(1/s)$

20. $\mathcal{L}\{1 - \cos 2t\} = (1/s) - s/(s^2 + 4)$

$\mathcal{L}\{(1 - \cos 2t)/t\} = \int_s^\infty [(1/s) - s/(s^2 + 4)]ds$

$= [\ln s/\sqrt{(s^2 + 4)}]_s^\infty$

$$= \ln \left[(1/s) \sqrt{(s^2 + 4)} \right]$$

21. $\mathcal{L}\{e^{3t} - 1\} = 1/(s - 3) - 1/s$

$\int [1/(s - 3) - 1/s]ds = \ln[(s - 3)/s] + C$

$\mathcal{L}[(e^{3t} - 1)/t] = \ln[s/(s - 3)]$

22. $\mathcal{L}\{e^t - e^{-t}\} = \mathcal{L}\{2 \sinh t\} = 2/(s^2 - 1)$

$\int [2/(s^2 - 1)]ds = \ln[(s - 1)/(s + 1)] + C$

$\mathcal{L}\{(e^t - e^{-t})/t\} = \ln[(s + 1)/(s - 1)]$

23. $f(t) = -(1/t)\mathcal{L}^1\{F'(s)\}$

$= -(1/t)\mathcal{L}^1\{1/(s - 2) - 1/(s + 2)\}$

$= -(1/t)(e^{2t} - e^{-2t}) = -(2 \sinh 2t)/t$

24. $f(t) = - (1/t)\mathcal{L}^1\{F'(s)\}$

$= - (1/t)\mathcal{L}^1\{2s/(s^2 + 1) - 2s/(s^2 + 4)\}$

$= (2/t)(\cos 2t - \cos t)$

25. $f(t) = -(1/t)\mathcal{L}^1\{2s/(s^2 + 1) - 1/(s + 2) - 1/(s - 3)\}$

$= (e^{-2t} + e^{3t} - 2 \cos t)/t$

26. $f(t) = - (1/t)\mathcal{L}^1\{F'(s)\}$

$= - (1/t)\mathcal{L}^1\{- 3/[(s + 2)^2 + 9]\}$

$= t^{-1}e^{-2t}\sin 3t$

27. $f(t) = -(1/t)\mathcal{L}^1\{(-2/s^3)/(1 + 1/s^2)\}$

$= (2/t)\mathcal{L}^1\{1/(s^3 + s)\}$

$$= (2/t)\mathcal{L}^1\{1/s - s/(s^2 + 1)\}$$

$$= (2/t)(1 - \cos t)$$

28. An empirical approach works best with this one. We can
construct transforms with powers of $(s^2 + 1)$ in their
denominators by differentiating the transforms of sin t
and cos t. Thus,

$$\mathcal{L}\{t \sin t\} = -(d/ds)[1/(s^2 + 1)] = 2s/(s^2 + 1)^2,$$
$$\mathcal{L}\{t \cos t\} = -(d/ds)[s/(s^2 + 1)] = (s^2 - 1)/(s^2 + 1)^2,$$
$$\mathcal{L}\{t^2 \cos t\} = -(d/ds)[(s^2 - 1)/(s^2 + 1)^2]$$
$$= (2s^3 - 6s)/(s^2 + 1)^3.$$

From the first and last of these it follows readily that

$$\mathcal{L}^1\{s/(s^2 + 1)^3\} = (t \sin t - t^2 \cos t)/8.$$

Alternatively, one could work out the repeated convolution

$$\mathcal{L}^1\{s/(s^2 + 1)^3\} = (\cos t)*((\sin t)*(\sin t)).$$

29. $-[s^2X - x'(0)]' - [sX]' - 2[sX] + [X] = 0$

$s(s + 1)X'(s) + 4sX(s) = 0$ $\qquad$ (separable)

$X(s) = A/(s + 1)^4$

$x(t) = Ct^3e^{-t}$

30. $-[s^2X(s) - x'(0)]' - 3[s X(s)]' - [s X(s)] + 3X(s) = 0$

$-(s^2 + 3s)X(s) - 3s X(s) = 0$ $\quad$ (separable)

$X(s) = C/(s + 3)^3$

$x(t) = A\ t^2 e^{-3t}$

31. $- [s^2 X - x'(0)]' + 4[sX]' - [sX] - 4[X]' + 2X = 0$

$(s^2 - 4s + 4)X' + (3s - 6)X = 0$ (separable)

$(s - 2)X' + 3X = 0$

$X(s) = A/(s - 2)^3$

$x(t) = Ct^2 e^{-2t}$

32. $- [s^2 X(s) - x'(0)]' - 2[s\ X(s)]' - 2[s\ X(s)] - 2X(s) = 0$

$- (s^2 + 2s)X'(s) - (4s + 4)X(s) = 0$ (separable)

$X(s) = C/s^2(s + 2)^2 = A[1/s - 1/s^2 - 1/(s + 2) - 1/(s + 2)^2]$

$x(t) = A(1 - t - e^{-2t} - te^{-2t})$

33. $- [s^2 X - x(0)]' - 2[sX] - [X]' = 0$

$(s^2 + 1)X' + 4sX = 0$ (separable)

$X(s) = A/(s^2 + 1)^2$

$x(t) = C(\sin t - t \cos t)$

34. $- (s^2 + 4s + 13)X'(s) - (4s + 8)X(s) = 0$

$X(s) = C/(s^2 + 4s + 13)^2 = C/[(s + 2)^2 + 9]^2$

It now follows from Problem 25 in Section 4.2 that

$x(t) = Ae^{-2t}(\sin 3t - 3t \cos 3t)$

36. $s^2 X(s) + 4X(s) = F(s)$

$X(s) = (1/2)F(s) \cdot [2/(s^2 + 4)]$

$x(t) = (1/2)f(t) * (\sin 2t)$

38. $X(s) = (1/3) F(s) \cdot [3/(s^2 + 4s + 13)]$

$x(t) = (1/3) f(t) * (e^{-2t} \sin 3t)$

SECTION 4.5

PERIODIC AND PIECEWISE CONTINUOUS FORCING FUNCTIONS

1. $f(t) = (t - 3) u_3(t)$

2. $f(t) = (t - 1) u_1(t) - (t - 3) u_3(t)$

   Hence

   $f(t) = 0$ if $t < 1$;

   $f(t) = t - 1$ if $1 \le t < 3$;

   $f(t) = 2$ if $t \ge 3$.

3. $f(t) = e^{-2(t-1)} u_1(t)$

4. $f(t) = e^{t-1} u_1(t) - e^2 e^{t-2} u_2(t)$

   Hence

   $f(t) = 0$ if $t < 1$;

   $f(t) = e^{t-1}$ if $1 \le t < 2$;

   $f(t) = e^{t-1} - e^t$ if $t > 2$.

5. $f(t) = u_\pi(t) \sin(t - \pi) = - u_\pi(t) \sin t$

6. $f(t) = u_1(t) \cos \pi(t - 1)$

   Hence

   $f(t) = 0$ if $t < 1$;

$$f(t) = \cos \pi(t - 1) = - \cos \pi t \quad \text{if} \quad t \geq 1.$$

7. $f(t) = \sin t - u_{2\pi}(t)\sin(t - 2\pi) = [1 - u_{2\pi}(t)]\sin t$

8. $f(t) = \cos \pi t - u_2(t)\cos \pi(t - 2) = [1 - u_2(t)]\cos \pi t$

Hence

$$f(t) = \cos \pi t \quad \text{if } t < 2;$$

$$f(t) = 0 \quad \text{if} \quad t \geq 2.$$

9. $f(t) = \cos \pi t + u_3(t)\cos \pi(t - 3) = [1 - u_3(t)]\cos \pi t$

10. $f(t) = 2 \, u_{\pi}(t)\cos 2(t - \pi) - 2 \, u_{2\pi}(t)\cos 2(t - 2\pi)$

$$= 2[u_{\pi}(t) - u_{2\pi}(t)]\cos 2t$$

Hence

$$f(t) = 0 \quad \text{if} \quad t < \pi \quad \text{or} \quad t \geq 2\pi;$$

$$f(t) = 2 \cos 2t \quad \text{if} \quad \pi \leq t < 2\pi.$$

11. $f(t) = 2[1 - u_3(t)]; \quad F(s) = (2/s)(1 - e^{-3s})$

12. $f(t) = 3[u_1(t) - u_4(t)]; \quad F(s) = 3(e^s - e^{-4s})/s$

13. $f(t) = [1 - u_{2\pi}(t)]\sin t; \quad F(s) = (1 - e^{-2\pi s})/(s^2 + 1)$

14. $f(t) = [1 - u_2(t)]\cos \pi t = \cos \pi t - u_2(t)\cos \pi(t - 2)$

$F(s) = s(1 - e^{-2s})/(s^2 + \pi^2)$

15. $f(t) = [1 - u_{3\pi}(t)]\sin t = \sin t + u_{3\pi}(t)\sin(t - 3\pi)$

$F(s) = (1 + e^{-3\pi s})/(s^2 + 1)$

16. $f(t) = [u_\pi(t) - u_{2\pi}(t)]\sin 2t$

$\qquad = u_\pi(t)\sin 2(t - \pi) - u_{2\pi}(t)\sin 2(t - 2\pi)$

$\quad F(s) = 2(e^{-\pi s} - e^{-2\pi s})/(s^2 + 4)$

17. $f(t) = [u_2(t) - u_3(t)]\sin \pi t$

$\qquad = u_2(t)\sin \pi(t - 2) + u_3(t)\sin \pi(t - 3)$

$\quad F(s) = \pi(e^{-2s} + e^{-3s})/(s^2 + \pi^2)$

18. $f(t) = [u_3(t) - u_5(t)]\cos \pi t/2$

$\qquad = u_3(t)\sin \pi(t - 3)/2 + u_5(t)\sin \pi(t - 5)/2$

$\quad F(s) = 2\pi(e^{-3s} + e^{-5s})/(4s^2 + \pi^2)$

19. $f(t) = (t + 1)u_1(t);\quad F(s) = e^{-s}(1/s + 1/s^2)$

20. $f(t) = [1 - u_1(t)]t + u_1(t)$

$\qquad = t + u_1(t) - u_1(t)g(t - 1)$

where $g(t) = t + 1$. Hence

$$F(s) = 1/s^2 + e^{-s}/s - e^{-s}(1/s^2 + 1/s) = (1 - e^{-s})/s^2.$$

21. $f(t) = t[1 - u_1(t)] + (2 - t)[u_1(t) - u_2(t)]$

$\qquad = t + 2u_1(t) - 2u_2(t) - 2u_1(t)g(t - 1) + u_2(t)h(t - 2)$

where $g(t) = t + 1$ and $h(t) = t + 2$.

$$F(s) = 1/s^2 + 2e^{-s}/s - 2e^{-2s}/s - 2e^{-s}(1/s + 1/s^2)$$
$$- e^{-2s}(2/s + 1/s^2)$$

$$= (1 - 2e^{-s} + e^{-2s})/s^2$$

22. $f(t) = [u_1(t) - u_2(t)]t^3$

$$= u_1(t)g(t - 1) - u_2(t)h(t - 2)$$

where

$$g(t) = (t + 1)^3 = t^3 + 3t^2 + 3t + 1,$$

$$h(t) = (t + 2)^3 = t^3 + 6t^2 + 12t + 8.$$

It follows that

$$F(s) = e^{-s}G(s) - e^{-2s}H(s)$$

$$= [(s^3 + 3s^2 + 6s + 6)e^{-s}$$

$$- (8s^3 + 12s^2 + 12s + 6)e^{-2s}]/s^4.$$

23. Simply apply Formula (12) with $p = 1$ and $f(t) = 1$.

24. $f(t) = \cos kt$ is periodic with period $p = 2\pi/k$. Apply Formula (12) and use the integral formula

$$\int e^{at}\cos bt\, dt = e^{at}(a \cos bt + b \sin bt)/(a^2 + b^2) + C.$$

25. Apply Formula (12) with $p = 2a$ and $f(t) = 1$ if $0 \le t \le a$, $f(t) = 0$ if $a < t \le 2a$.

26. Apply Formula (12) with $f(t) = t/a$ and $p = a$; then use the integral formula $\int ue^u du = (u - 1)e^u + C$.

27. $G(s) = (1/as^2) - F(s)$. Now substitute for $F(s)$ the result of Problem 26.

28. $F(s) = (1 - e^{-as} - ase^{-as})/s^2(1 - e^{-2as})$

29. Apply Formula (12) with $p = 2\pi/k$ and $f(t) = \sin kt$ if $0 \le t \le \pi k$, $f(t) = 0$ if $\pi/k \le t \le 2\pi/k$; then use the integral formula

$$\int e^{at} \sin bt \, dt = e^{at}(a \sin bt - b \cos bt)/(a^2 + b^2) + C.$$

30. $h(t) = f(t) + u(t - \pi/k) \, f(t - \pi/k),$ so

$$
\begin{aligned}
H(s) &= F(s) + e^{-\pi s/k}F(s) = (1 + e^{-\pi s/k})F(s) \\
&= (1 + e^{-\pi s/k}) \cdot k/(s^2 + k^2)(1 - e^{-\pi s/k}) \\
&= [k/(s^2 + k^2)] \cdot [(1 + e^{-\pi s/k})/(1 + e^{-\pi s/k})] \\
&= [k/(s^2 + k^2)]\coth \pi s/2k
\end{aligned}
$$

31. $x'' + 4x = 1 - u_\pi(t)$

$s^2X(s) + 4X(s) = (1 - e^{-\pi s})/s$

$$
\begin{aligned}
X(s) &= (1 - e^{-\pi s})/s(s^2 + 4) \\
&= (1/4)(1 - e^{-\pi s})[1/s - s/(s^2 + 4)]
\end{aligned}
$$

$$
\begin{aligned}
x(t) &= (1/4)[1 - u_\pi(t)][1 - \cos 2(t - \pi)] \\
&= (1/2)[1 - u_\pi(t)]\sin^2 t
\end{aligned}
$$

32. $x'' + 5x' + 4x = 1 - u_2(t)$

$s^2X(s) + 5s \, X(s) + 4X(s) = (1 - e^{-2s})/s$

$$X(s) = (1 - e^{-2s})/s(s^2 + 5s + 4) = (1 - e^{-2s})G(s)$$

where

$$g(t) = (1/4)(1 - e^{-2t}\cos t - 3e^{-2t}\sin t)$$

It follows that

$$x(t) = g(t) - u_2(t)g(t - 2).$$

Hence

$$x(t) = g(t) \quad \text{if} \quad t < 2, \quad \text{and}$$

$$x(t) = g(t) - g(t - 2) \quad \text{if} \quad t \geq 2.$$

33. $x'' + 9x = [1 - u_{2\pi}(t)]\sin t$

$X(s) = (1 - e^{-2\pi s})/[(s^2 + 1)(s^2 + 9)]$

$\qquad = (1/8)(1 - e^{-2\pi s})[1/(s^2 + 1) - 1/(s^2 + 9)]$

$x(t) = (1/8)[1 - u_{2\pi}(t)][\sin t - (1/3)\sin 3t]$

34. $x'' + x = [1 - u_1(t)]t$

$s^2 X(s) + X(s) = 1/s^2 - e^{-s}(1/s + 1/s^2)$

It follows that

$$X(s) = \frac{1}{s^2(s^2 + 1)} - \frac{e^{-s}(s^2 + 1)}{s^2(s^2 + 1)}$$

$$\qquad = (1 - e^{-s}) \left[\frac{1}{s^2} - \frac{1}{s^2 + 1} \right] - e^{-s} \left[\frac{1}{s} - \frac{s}{s^2 + 1} \right]$$

$$X(s) = (1 - e^{-s})G(s) - e^{-s}H(s)$$

where $g(t) = t - \sin t$, $h(t) = 1 - \cos t$. Hence

$$x(t) = g(t) - u_1(t)g(t - 1) - u_1(t)h(t - 1)$$

and so

$\qquad x(t) = t - \sin t \quad \text{if } t < 1,$

$\qquad x(t) = -\sin t + \sin(t - 1) + \cos(t - 1) \quad \text{if} \quad t > 1.$

35. $x'' + 4x' + 4x = [1 - u_2(t)]t = t - u_2(t)[(t - 2) + 2]$

$(s + 2)^2 X(s) = 1/s^2 - e^{-2s}(2/s + 1/s^2)$

$$X(s) = [1/s^2(s + 2)^2] - e^{-2s}[(2s + 1)/s^2(s + 2)^2]$$

$$\qquad = (1/4)[-1/s + 1/s^2 + 1/(s + 2) + 1/(s + 2)^2]$$

$$- (1/4) e^{-2s} [1/s + 1/s^2 - 1/(s + 2) - 3/(s + 2)^2]$$

$$x(t) = (1/4) \{-1 + t + (1 + t) e^{-2t}$$
$$+ u_2(t) [1 - t + (3t - 5) e^{-2(t-2)}]\}$$

36. $100 \ I(s) + 1000 \ I(s)/s = 100(1/s - e^{-s}/s)$

$I(s) = (1 - e^{-s})/(s + 10) = (1 - e^{-s}) G(s)$

where $g(t) = e^{-10t}$. Hence

$$i(t) = e^{-10t} - u_1(t) e^{-10(t-1)}$$

37. $i'(t) + 10^4 \int i(t) dt = 100 [1 - u_{2\pi}(t)]$

$sI(s) + 10^4 I(s)/s = 100(1 - e^{-2\pi s})/s$

$$I(s) = 100(1 - e^{-2\pi s})/(s^2 + 10^4)$$

$$i(t) = \sin 100t - u_{2\pi}(t) \sin 100(t - 2\pi)$$
$$= [1 - u_{2\pi}(t)] \sin 100t$$

38. $i'(t) + 10000 \int i(t) dt = [1 - u_\pi(t)] (100 \sin 10t)$

$s \ I(s) + 10000 \ I(s)/s = 1000(1 - e^{-\pi s})/(s^2 + 100)$

Hence

$$I(s) = (1 - e^{-\pi s}) \cdot 1000s/(s^2 + 100)(s^2 + 10000)$$
$$= (1 - e^{-\pi s}) G(s)$$

with

$$g(t) = (10/99)(\cos 10t - \cos 100t).$$

It follows that

$$i(t) = g(t) - u_\pi(t)g(t - \pi),$$

so

$$i(t) = g(t) \quad \text{if} \quad t < \pi, \quad i(t) = 0 \quad \text{if} \quad t > \pi.$$

39. $i'(t) + 150\ i(t) + 5000 \int i(t)dt = 100t[1 - u_1(t)]$

$sI(s) + 150I(s) + 5000I(s)/s = 100/s^2 - 100e^{-s}(1/s + 1/s^2)$

$I(s) = [100/s(s + 50)(s + 100)]$

$$- e^{-s}[100(s + 1)/s(s + 50)(s + 100)]$$

$= (1/50)[1/s - 2/(s + 50) + 1/(s + 100)]$

$$- (1/50)e^{-s}[1/s + 98/(s + 50) - 99/(s + 100)]$$

$i(t) = (1/50)[1 - 2e^{-50t} + e^{-100t}]$

$$- (1/50)u_1(t)[1 + 98e^{-50(t-1)} - 99e^{-100(t-1)}]$$

40. $i'(t) + 100\ i(t) + 2500 \int i(t)dt = 50t[1 - u_1(t)]$

$s\ I(s) + 100\ I(s) + 2500\ I(s)/s = 50/s^2 - 50e^{-s}(1/s + 1/s^2)$

It follows that

$$I(s) = (1 - e^{-s})[50/s(s + 50)^2] - e^{-s}[50/(s + 50)^2]$$

$$= (1 - e^{-s})G(s) - e^{-s}H(s)$$

where

$$g(t) = (1 - e^{-50t} - 50te^{-50t})/50,$$

$$h(t) = 50te^{-50t}.$$

Hence

$$i(t) = g(t) - u_1(t)g(t - 1) - u_1(t)h(t - 1).$$

41. $x'' + 4x = f(t), \quad x(0) = x'(0) = 0$

$(s^2 + 4)X(s) = (4s)[(1 - e^{-\pi s})/(1 + e^{-\pi s})]$ (by Example 6)

By a computation like that in the solution of Example 8 it follows that

$$(s^2 + 4)X(s) = (4/s) + (8/s) \sum (-1)^n e^{-n\pi s}.$$

Now let

$$g(t) = \mathcal{L}^{-1}\{4/s(s^2 + 4)\}$$
$$= 1 - \cos 2t = 2 \sin^2 t.$$

Then it follows that

$$x(t) = g(t) + 2 \sum (-1)^n u_{n\pi}(t)g(t - n\pi)$$
$$= 2 \sin^2 t + 4 \sum (-1)^n u_{n\pi}(t)\sin^2 t.$$

Hence

$$x(t) = 2 \sin^2 t \quad \text{if} \quad 2n\pi \le t < (2n + 1)\pi,$$
$$x(t) = -2 \sin^2 t \quad \text{if} \quad (2n - 1)\pi \le t < 2n\pi.$$

Consequently the complete solution

$$x(t) = 2|\sin t|\sin t$$

is periodic, so the transient solution is zero.

42. $x'' + 2x' + 10x = f(t)$, $x(0) = x'(0) = 0$

As in the solution of Example 8 we find first that

$$(s^2 + 2s + 10)X(s) = (10/s) + (20/s) \sum (-1)^n e^{-n\pi s}.$$

If

$$g(t) = \mathcal{L}^1\{10/s[(s + 1)^2 + 9]\}$$

$$= 1 - (1/3)e^{-t}(3 \cos 3t + \sin 3t)$$

then it follows that

$$x(t) = g(t) + 2 \sum (-1)^n u_{n\pi}(t)g(t - n\pi).$$

SECTION 4.6

IMPULSES AND DELTA FUNCTIONS

Among the several ways of introducing delta functions, we consider the physical approach of the first two pages of this section to be the most tangible one for elementary students. Whatever the approach, however, the practical consequences are the same -- as described in the discussion associated with equations (11)-(19). That is, in order to solve a differential equation of the form

$$ax''(t) + bx'(t) + cx(t) = f(t)$$

where $f(t)$ involves delta functions, we transform the equation using the operational principle $\mathcal{L}\{\delta_a(t)\} = e^{-as}$, solve for $X(s)$, and finally invert as usual to find the formal solution $x(t)$.

1. $s^2X(s) + 4X(s) = 1$

$$X(s) = 1/(s^2 + 4)$$

$$x(t) = (1/2)\sin 2t$$

2. $s^2X(s) + 4X(s) = 1 + e^{-\pi s}$

   $X(s) = 1/(s^2 + 4) + e^{-\pi s}/(s^2 + 4)$

   $x(t) = (1/2)\sin 2t + (1/2)u_\pi(t)\sin 2t$

Hence

   $x(t) = (1/2)\sin 2t \quad \text{if} \quad t < \pi$

   $x(t) = \qquad \sin 2t \quad \text{if} \quad t > \pi.$

3. $s^2X(s) + 4sX(s) + 4X(S) = (1/s) + e^{-2s}$

   $X(s) = [1/s(s + 2)^2] + e^{-2s}/(s + 2)^2$

   $= (1/4)[1/s - 1/(s + 2) - 2/(s + 2)^2] + e^{-2s}/(s + 2)^2$

   $x(t) = (1/4)[1 - e^{-2t} - 2te^{-2t}] + u_2(t)(t - 2)e^{-2(t-2)}$

4. $[s^2X(s) - 1] + 2sX(s) + X(s) = 1 + 1/s^2$

   $X(s) = (2s^2 + 1)/s^2(s + 1)^2$

   $= -2/s + 1/s^2 + 2/(s + 1) + 3/(s + 1)^2$

   $x(t) = -2 + t + 2e^{-t} + 3t\,e^{-t}$

5. $(s^2 + 2s + 2)X(s) = 2e^{-\pi s}$

   $X(s) = 2e^{-\pi s}/[(s + 1)^2 + 1]$

   $x(t) = 2u_\pi(t)e^{-(t-\pi)}\sin(t - \pi)$

Hence

   $x(t) = 0 \quad \text{if} \quad t \le \pi,$

   $x(t) = -2e^{-(t-\pi)}\sin t \quad \text{if} \quad t \ge \pi.$

6. $s^2X(s) + 9X(s) = e^{-3\pi s} + s/(s^2 + 9)$

$X(s) = e^{-3\pi s}/(s^2 + 9) + s/(s^2 + 9)^2$

$x(t) = u_{3\pi}(t) \cdot (1/3)\sin 3(t - 3\pi) + (1/6)t \sin 3t$
$= -(1/3)u_{3\pi}(t)\sin 3t + (1/6)t \sin t$

7. $[s^2X(s) - 2] + 4sX(s) + 5X(s) = e^{-\pi s} + e^{-2\pi s}$

$X(s) = [2 + e^{-\pi s} + e^{-2\pi s}]/[(s + 2)^2 + 1]$

$x(t) = 2e^{-2t}\sin t + u_{\pi}(t)e^{-2(t-\pi)}\sin(t - \pi)$
$+ u_{2\pi}(t)e^{-2(t-2\pi)}\sin(t - 2\pi)$

$x(t) = [2 - e^{2\pi}u_{\pi}(t) + e^{4\pi}u_{2\pi}(t)]e^{-2t}\sin t$

8. $[s^2X(s) - 2s - 2] + 2[sX(s) - 2] + X(s) = 1 - e^{-2s}$

$X(s) = (2s + 7 - e^{-2s})/(s + 1)^2$
$= 2/(s + 1) + 5/(s + 1)^2 - e^{-2s}/(s + 1)^2$

$x(t) = (2 + 5t)e^{-t} - u_2(t)(t - 2)e^{-(t-2)}$

9. $s^2X(s) + 4X(s) = F(s)$

$X(s) = [1/(s^2 + 4)] \cdot F(s)$

$x(t) = \frac{1}{2} \int_0^t (\sin 2u)f(t - u)du$

10. $s^2X(s) + 6s\,X(s) + 9X(s) = F(s)$

$\qquad X(s) = [1/(s + 3)^2]\cdot F(s)$

$\qquad x(t) = \displaystyle\int_0^t u\,e^{-3u}f(t - u)\,du$

11. $(s^2 + 6s + 8)X(s) = F(s)$

$\qquad X(s) = \{1/[(s + 3)^2 - 1]\}\cdot F(s)$

$\qquad x(t) = \displaystyle\int_0^t e^{-3u}(\sinh u)\,f(t - u)\,du$

12. $s^2X(s) + 4sX(s) + 8X(s) = F(s)$

$\qquad X(s) = \{1/[(s + 2)^2 + 4]\}\cdot F(s)$

$\qquad x(t) = \dfrac{1}{2}\displaystyle\int_0^t (e^{-2u}\sin 2u)\,f(t - u)\,du$

13. (a) $mx''(t) = (p/\epsilon)[u_0(t) - u_\epsilon(t)]$

$\quad ms^2X_\epsilon(s) = (p/\epsilon)\,[1/s - e^{-\epsilon s}/s]$

$\quad mX_\epsilon(s) = (p/\epsilon)[(1 - e^{-\epsilon s})/s^3]$

$\quad mx_\epsilon(t) = (p/2\epsilon)[t^2 - u_\epsilon(t)(t - \epsilon)^2]$

(b) If $t > \epsilon$ then

$\quad mx_\epsilon(t) = (p/2\epsilon)[t^2 - (t^2 - 2\epsilon t + \epsilon^2)]$
$\qquad\qquad = (p/2\epsilon)(2\epsilon t - \epsilon^2).$

Hence $mx_\epsilon(t) \longrightarrow pt$ as $\epsilon \longrightarrow 0.$

(c) $mv = (mx)' = (pt)' = p.$

14. $sX(s) = e^{-as}$

$X(s) = e^{-as}/s$

$x(t) = u(t - a)$

15. Each of the two given initial value problems transforms to

$$(ms^2 + k)X(s) = mv_0 = p_0.$$

16. Each of the two given initial value problems transforms to

$$(as^2 + bs + c)X(s) = F(s) + av_0$$

17. (b) $i' + 100i = \delta_1(t) - \delta_2(t), \quad i(0) = 0$

$I(s) = (e^{-s} - e^{-2s})/(s + 100)$

$i(t) = u_1(t)e^{-100(t-1)} - u_2(t)e^{-100(t-2)}$

$\quad = e^{-100t}[e^{100}u_1(t) - e^{200}u_2(t)]$

18. (b) $i''(t) + 100\ i(t) = 10\ \delta(t) - 10\ \delta(t - \pi)$

$(s^2 + 100)\ I(s) = 10 - 10\ e^{-\pi s}$

$I(s) = 10/(s^2 + 100) - 10\ e^{-\pi s}/(s^2 + 100)$

$i(t) = \sin 10t - u_\pi(t)\sin 10(t - \pi)$

$\quad = [1 - u_\pi(t)]\sin 10t$

19. $(s^2 + 100)I(s) = 10 \sum (-1)^n e^{-n\pi s/10}$

$I(s) = \sum (-1)^n e^{-n\pi s/10}[10/(s^2 + 100)]$

$$i(t) = \sum (-1)^n u_{n\pi/10}(t) \sin(10t - n\pi)$$
$$= \sum u_{n\pi/10}(t) \sin 10t$$

because $\sin(10t - n\pi) = (-1)^n \sin 10t$. Hence

$$i(t) = (n + 1) \sin 10t$$

if $n\pi/10 < t < (n + 1)\pi/10$.

20. $(s^2 + 100) I(s) = 10 \sum (-1)^n e^{-n\pi s/5}$

$$I(s) = \sum (-1)^n e^{-n\pi s/5} [10/(s^2 + 100)]$$

$$i(t) = \sum (-1)^n u_{n\pi/5}(t) \sin(10t - 2n\pi)$$
$$= \sum (-1)^n u_{n\pi/5}(t) \sin 10t.$$

Hence

$$i(t) = \sin 10t - \sin 10t + \cdots + (-1)^n \sin 10t$$

if $n\pi/5 < (n + 1)\pi/5, \quad n \geq 0$.

21. (b) $(s^2 + 60s + 1000) I(s) = 10 \sum (-1)^n e^{-n\pi s/10}$

where the summation is from $n = 0$ to ∞. Hence

$$I(s) = \sum (-1)^n 10 \, e^{-n\pi s/10}/[(s + 30)^2 + 100]$$
$$i(t) = \sum (-1)^n u_{n\pi/10}(t) \, g(t - n\pi/10)$$

where $g(t) = e^{-30t} \sin 10t$, and so

$$g(t - n\pi/10) = \exp[- 30(t - n\pi/10)] \sin 10(t - n\pi/10)$$
$$= e^{3n\pi}e^{-30t} \cdot (-1)^n \sin 10t$$

Therefore

$$i(t) = \sum u_{n\pi/10}(t) \ e^{3n\pi}e^{-30t}\sin 10t.$$

If $\quad n\pi/10 < t < (n + 1)\pi/10 \quad$ then it follows that

$$i(t) = (1 + e^{3\pi} + \cdots + e^{3n\pi})e^{-30t}\sin 10t$$
$$= [(e^{3n\pi+3\pi} - 1)/(e^{3\pi} - 1)]e^{-30t}\sin 10t.$$

22. $(s^2 + 1)X(s) = \sum e^{-2n\pi s}$

$$X(s) = \sum e^{-2n\pi s}[1/(s^2 + 1)]$$

$$x(t) = \sum u_{2n\pi}(t)\sin(t - 2n\pi)$$
$$= \sum u_{2n\pi}(t)\sin t$$

Hence $\quad x(t) = (n + 1)\sin t \quad$ if $\quad 2n\pi < t < 2(n + 1)\pi.$

CHAPTER 5

LINEAR SYSTEMS OF DIFFERENTIAL EQUATIONS

This chapter is designed to offer considerable flexibility in the treatment of linear systems, depending on the background in linear algebra that students are assumed to have. Sections 5.1 and 5.2 can stand alone as an introduction to linear systems without the use of linear algebra and matrices. The remaining sections of Chapter 5 employ the notation and terminology (though not so much theory) of elementary linear algebra. For ready reference and review, Section 5.3 includes a complete and self-contained account of the needed background of determinants, matrices, and vectors.

SECTION 5.1

INTRODUCTION TO SYSTEMS

1. $x_1' = x_2$ $\quad (x = x_1)$
    $x_2' = -7x_1 - 3x_2 + t^2$

2. $x_1' = x_2,$ $\quad x_2' = x_3,$ $\quad x_3' = x_4$ $\quad (x = x_1)$
    $x_4' = -x_1 + 3x_2 - 6x_3 + \cos 3t$

3. $x_1' = x_2$ $\quad (x = x_1)$
    $t^2 x_2' = (1 - t^2)x_1 - tx_2$

4. $x_1' = x_2,$ $\qquad x_2' = x_3$ $\qquad (x = x_1)$

$t^3 x_3' = -5x_1 - 3tx_2 + 2t^2 x_3 + \ln t$

5. $x_1' = x_2,$ $\qquad x_2' = x_3$ $\qquad (x = x_1)$

$x_3' = x_2^2 + \cos x_1$

6. $x_1' = x_2$ $\qquad (x = x_1)$

$y_1' = y_2$ $\qquad (y = y_1)$

$x_2' = 5x_1 - 4y_1$

$y_2' = -4x_1 + 5y_1$

7. $x_1' = x_2$ $\qquad (x = x_1)$

$y_1' = y_2$ $\qquad (y = y_1)$

$x_2' = -kx_1(x_1^2 + y_1^2)^{-3/2}$

$y_2' = -ky_1(x_1^2 + y_1^2)^{-3/2}$

8. $x_1' = x_2$ $\qquad (x = x_1)$

$y_1' = y_2$ $\qquad (y = y_1)$

$x_2' = -4x_1 + 2y_1 - 3x_2$

$y_2' = 3x_1 - y_1 - 2y_2 + \cos t$

9. $x_1' = x_2$ $\qquad (x = x_1)$

$y_1' = y_2$ $\qquad (y = y_1)$

$z_1' = z_2$ $\qquad (z = z_1)$

$x_2' = 3x_1 - y_1 + 2z_1$

$y_2' = x_1 + y_1 - 4z_1$

$z_2' = 5x_1 - y_1 - z_1$

10. $x_1' = x_2,$ $y_1' = y_2$

    $x_2' = x_1(1 - y_1)$

    $y_2' = y_1(1 - x_1)$

11. Because of the sin 2t term in the first equation we try

$$x = A \sin 2t, \quad y = B \sin 2t.$$

Substitution in the two differential equations, followed by cancellation of sin 2t throughout, yields the system

$$A - 3B = 1, \quad -3A + B = 0$$

that we solve readily for $A = -1/8$ and $B = -3/8$. Thus we get the particular solution

$$x = -(1/8)\sin 2t, \quad y = -(3/8)\sin 2t.$$

12. $x = (8/105)\cos 4t,$ $y = (-22/105)\cos 4t$

13. $x = -(3/8)\cos 3t,$ $y = (1/4)\cos 3t$

14. The transformed equations are

$$s\, X(s) - 1 = X(s) + 2Y(s),$$
$$s\, Y(s) \quad = 2X(s) + Y(s),$$

which we solve for

$$X(s) = (s - 1)(s^2 - 2s - 3) = (1/2)[1/(s - 3) + 1/(s + 1)],$$
$$Y(s) = \quad 2/(s^2 - 2s - 3) = (1/2)[1/(s - 3) - 1/(s + 1)].$$

Hence the solution is

$$x(t) = (e^{3t} + e^{-t})/2, \quad y(t) = (e^{3t} - e^{-t})/2.$$

15. $sX(s) - 1 = 2X(s) + Y(s)$
$sY(s) + 2 = 6X(s) + 3Y(s)$

$X(s) = (s - 5)/[s(s - 5)] = 1/s$
$Y(s) = (-2s + 10)/[s(s - 5)] = -2/s$

$x(t) = 1, \quad y(t) = -2$

16. The transformed equations are

$s\,X(s) = X(s) + 2Y(s),$
$s\,Y(s) = X(s) + 1/(s + 1),$

which we solve for

$X(s) = 2/(s - 2)(s + 1)^2$
$\quad = (2/9)[1/(s - 2) - 1/(s + 1) - 3/(s + 1)^2],$

$Y(s) = (s - 1)/(s - 2)(s + 1)^2$
$\quad = (1/9)[1/(s - 2) - 1/(s + 1) + 6/(s + 1)^2].$

Hence the solution is

$x(t) = (2/9)(e^{2t} - e^{-t} - 3t\,e^{-t}),$
$y(t) = (1/9)(e^{2t} - e^{-t} + 6t\,e^{-t}).$

17. $sX(s) + 2[sY(s) - 1] + X(s) = 0$
$sX(s) - [sY(s) - 1] + Y(s) = 0$

$X(s) = -2/(3s^2 - 1)$
$Y(s) = (3s + 1)/(3s^2 - 1)$

$x(t) = -(2/\sqrt{3})\sinh(t/\sqrt{3})$
$y(t) = \cosh(t/\sqrt{3}) + (1/\sqrt{3})\sinh(t/\sqrt{3})$

18. The transformed equations are

$s^2X(s) + 1 + 2X(s) + 4Y(s) = 0,$

$s^2Y(s) + 1 + X(s) + 2Y(s) = 0,$

which we solve for

$X(s) = (-s^2 + 2)/s^2(s^2 + 4) = (1/4)[2/s^2 - 6/(s^2 + 4)],$

$Y(s) = (-s^2 - 1)/s^2(s^2 + 4) = (-1/8)[2/s^2 + 6/(s^2 + 4)].$

Hence the solution is

$x(t) = (1/4)(2t - 3 \sin 2t),$

$y(t) = (-1/8)(2t + 3 \sin 2t).$

19. $[s^2X - s] + [sX - 1] + [sY - 1] + 2X - Y = 0$

$[s^2Y - s] + [sX - 1] + [sY - 1] + 4X - 2Y = 0$

$X(s) = (2s^2 - 2)/s(s^3 + 2s^2 - 3) = (2s + 2)/s(s^2 + 3s + 3)$

$= (1/3)[2/s + (s + 3)/(s^2 + 3s + 3)]$

$Y(s) = (s^3 + 2s^2 - 2s - 4)/s(s - 1)(s^2 + 3s + 3)$

$= (1/21)[28/s - 9/(s - 1) + (2s + 15)/(s^2 + 3s + 3)]$

$x(t) = (1/3)\{2 + e^{-3t/2}[\cos(t/2)\sqrt{3} + \sqrt{3} \sin(t/2)\sqrt{3}]\}$

$y(t) = (1/21)\{28 - 9e^t$
$\qquad\qquad + 2e^{-3t/2}[\cos(t/2)\sqrt{3} + 4\sqrt{3} \sin(t/2)\sqrt{3}]\}$

20. The transformed equations are

$s X(s) - 1 = X(s) + Z(s),$

$s Y(s) = X(s) + Y(s),$

$s Z(s) = -2X(s) - Z(s),$

which we solve for

$X(s) = (s^2 - 1)/(s - 1)(s^2 + 1) = (s + 1)/(s^2 + 1),$

$Y(s) = (s + 1)/(s - 1)(s^2 + 1) = 1/(s - 1) - s/(s^2 + 1),$

$Z(s) = -2(s - 1)/(s - 1)(s^2 + 1) = -2/(s^2 + 1).$

Hence the solution is

$$x(t) = \cos t + \sin t,$$
$$y(t) = e^t - \cos t,$$
$$z(t) = -2 \sin t.$$

21. $20[sX(s) - 50] + 6X(s) - Y(s) = 0$
     $20[sY(s) - 50] - 6X(s) + 3Y(s) = 0$

$$X(s) = 1000(5s + 1)/(100s^2 - 45s + 3)$$
$$Y(s) = 1000(5s + 3)/(100s^2 + 45s + 3)$$

Let $a = (-9 - \sqrt{33})/40$, $b = (-9 + \sqrt{33})/40$. Then

$$x(t) = 25(1 + 1/\sqrt{33})e^{at} + 25(1 - 1/\sqrt{33})e^{bt},$$
$$y(t) = 25(1 - 15/\sqrt{33})e^{at} + 25(1 + 15/\sqrt{33})e^{bt}.$$

22. Using small letters for currents and capital letters for
     transforms, we want to solve the system

$$i_1'(t) + 25\ i_1(t) - 25\ i_2(t) = 50,$$
$$2i_1'(t) - 3\ i_2'(t) - 5\ i_2(t) = 0$$

with initial conditions $i_1(0) = i_2(0) = 0$. The transformed
system is

$$s\ I_1(s) + 25\ I_1(s) - 25\ I_2(s) = 50/s,$$
$$2s\ I_1(s) - 3s\ I_2(s) - 5\ I_2(s) = 0.$$

We solve for

$$I_1(s) = 50(3s + 5)/\{3s[(s + 5)^2 + 50/3]\},$$
$$I_2(s) = 100/\{3[(s + 5)^2 + 50/3]\}.$$

Using the given inverse Laplace transform, we find finally
that

$$i_1(t) = 2 + e^{-5t}[4\sqrt{6} \sin t\sqrt{(50/3)} - 2 \cos t\sqrt{(50/3)}],$$
$$i_2(t) = (20/\sqrt{6})e^{-5t}\sin t\sqrt{(50/3)}.$$

29. $2(I_1' - I_2') + 50I_1 = 100 \sin 60t$

    $2(I_2' - I_1') + 25I_2 = 0$

30. First we apply Kirchoff's law to each loop in Figure 5.9, denoting by Q the charge on the capacitor, and get the equations

    $$50 \ I_1 + 1000 \ Q = 100, \quad 25 \ I_2 - 1000 \ Q = 0.$$

    Then we differentiate each equation and substitute $Q' = I_1 - I_2$ to get the system

    $$I_1' = - \ 20(I_1 - I_2), \quad I_2' = 40(I_1 - I_2).$$

32. If we write (x',y') for the velocity vector and

    $$v = [(x')^2 + (y')^2]^{1/2}$$

    for the speed, then $(x'/v, y'/v)$ is a unit vector pointing in the direction of the velocity vector, and so the components of the air resistance force F_r are given by

    $$F_r = - \ kv^2(x'/v, \ y'/v) = (- \ kvx', \ - \ kvy').$$

SECTION 5.2

THE METHOD OF ELIMINATION

1. $x = a_1 e^{-t} + a_2 e^{2t}, \quad y = a_2 e^{2t}$

2. From the first differential equation we get $y = (x - x')/2$, so $y' = (x' - x'')/2$. Substitution of these expressions for y and y' into the second differential equation yields the second order equation

    $$x'' + 2x' + x = 0,$$

with general solution

$$x = (c_1 + c_2 t) e^{-t}.$$

Substitution in $y = (x - x')/2$ now yields

$$y = (c_1 - c_2/2 + c_2 t) e^{-t}.$$

3. $x = 4(e^{3t} - e^{-2t})/5, \quad y = 2(6e^{3t} - e^{-2t})/5$

4. Elimination of y and y' yields the second order equation $x'' - 4x = 0$ with general solution

$$x = c_1 e^{2t} + c_2 e^{-2t}.$$

Substitution in $y = 3x - x'$ gives

$$y = c_1 e^{2t} + 5c_2 e^{-2t}.$$

The initial conditions yield the equations

$$c_1 + c_2 = 1, \quad c_1 + 5c_2 = -1$$

with solution $c_1 = 3/2$, $c_2 = -1/2$. Hence the desired particular solution is

$$x = (3e^{2t} - e^{-2t})/2, \quad y = (3e^{2t} - 5e^{-2t})/2.$$

5. $x = e^{-t}(a_1 \cos 2t + a_2 \sin 2t)$

$y = -(1/2) e^{-t}[(a_1 + a_2) \cos 2t + (a_2 - a_1) \sin 2t]$

6. General solution:

$$x = e^{-2t}(c_1 \cos 3t + c_2 \sin 3t)$$

$$y = (1/3) e^{-2t}[(-c_1 + c_2) \cos 3t - (c_1 + c_2) \sin 3t]$$

Particular solution:

$$x = e^{-2t}(3 \cos 3t + 9 \sin 3t)$$

$$y = e^{-2t}(2 \cos 3t - 4 \sin 3t)$$

7. $x = a_1 e^{2t} + a_2 e^{3t} - t/3 + 1/8$

    $y = -2a_1 e^{2t} - a_2 e^{3t} - 2t/3 - 5/9$

8. $x = c_1 e^t + c_2 e^{3t} + e^{2t}, \qquad y = -c_1 e^t + c_2 e^{3t}$

9. $x = 3a_1 e^t + a_2 e^{-t} - (1/5)[7 \cos 2t + 4 \sin 2t]$

    $y = a_1 e^t - a_2 e^{-t} - (1/5)[2 \cos 2t + 4 \sin 2t]$

10. First we solve the given equations for

    $$x' = 2x + y, \qquad y' = x + 2y.$$

    The general solution is

    $$x = c_1 e^t + c_2 e^{3t}, \qquad y = -c_1 e^t + c_2 e^{3t}.$$

    The desired particular solution is

    $$x = e^t, \quad y = -e^t.$$

11. $x = a_1 \cos 3t + a_2 \sin 3t - (11/20)e^t - (1/4)e^{-t}$

    $y = (1/3)[(a_1 - a_2)\cos 3t + (a_1 + a_2)\sin 3t] + (1/10)e^t$

12. The first equation yields $y = (x'' - 6x)/2$, so
    $y'' = (x^{(4)} - 6x'')/2$. Substitution in the second equation
    yields

    $$x^{(4)} - 13x'' + 36x = 0.$$

    The general solution is

$$x = c_1 e^{2t} + c_2 e^{-2t} + \qquad c_3 e^{3t} + \qquad c_4 e^{-3t},$$
$$y = - c_1 e^{2t} - c_2 e^{-2t} + (3/2) c_3 e^{3t} + (3/2) c_4 e^{-3t}.$$

13. $x = a_1 \cos 2t + a_2 \sin 2t \quad + \quad b_1 \cos 3t + b_2 \sin 3t$
     $y = (a_1 \cos 2t + a_2 \sin 2t)/2 - 2(b_1 \cos 3t + b_2 \sin 3t)$

14. $x = c_1 \cos 2t + c_2 \sin 2t + (1/3) \sin t$
     $y = c_1 \cos 2t + c_2 \sin 2t$
     $\qquad + c_3 \cos 2t\sqrt{2} + c_4 \sin 2t\sqrt{2} + (4/21) \sin t$

15. $x = a_1 \cos t + a_2 \sin t + b_1 \cos 2t + b_2 \sin 2t$
     $y = a_2 \cos t - a_1 \sin t + b_2 \cos 2t - b_1 \sin 2t$

16. In operational form our system is

$$(D^2 - 4)x + \qquad 13Dy = 6 \sin t$$
$$- 2Dx + (D^2 - 9)y = 0$$

When we operate on the first equation with $D^2 - 9$, on the second with $13D$, and subtract, the result is

$$(D^4 + 13D^2 + 36)x = - 60 \sin t.$$

The general solution of this fourth order equation is

$$x = a_1 \cos 2t + a_2 \sin 2t + a_3 \cos 3t + a_4 \sin 3t - (5/2) \sin t.$$

We find similarly that

$$y = b_1 \cos 2t + b_2 \sin 2t + b_3 \cos 3t + b_4 \sin 3t + (1/2) \cos t.$$

When we substitute these expressions into either of the original differential equations we find that

$$b_1 = - 4a_2/13, \quad b_2 = 4a_1/13, \quad b_3 = - a_4/3, \quad b_4 = a_3/3.$$

Hence

$$y = (-4a_2\cos 2t + 4a_1\sin 2t)/13$$
$$+ (-a_4\cos 3t + a_3\sin 3t)/3 + (1/2)\cos t.$$

17. $x = a_1\cos t + a_2\sin t + b_1e^{2t} + b_2e^{-2t}$
    $y = 3a_2\cos t + 3a_1\sin t - b_1e^{2t} + b_2e^{-2t}$

18. From the first and third equations we see that $x' + z' = 0$, so $z = -x$. Hence the first equation reduces to $x' = 2y$, and substitution in the second equation yields $y'' + y' - 12y = 0$, so
    $$y = c_1e^{3t} + c_2e^{-4t}.$$

    Since $x = (y + y')/6$, it follows that
    $$x = (4c_1e^{3t} - 3c_2e^{-4t})/6,$$
    $$z = (-4c_1e^{3t} + 3c_2e^{-4t})/6.$$

19. $x = a_1 + a_2e^{4t} + a_3e^{8t}$
    $y = 2a_1 \qquad\qquad - 2a_3e^{8t}$
    $z = 2a_1 - 2a_2e^{4t} + 2a_3e^{8t}$

20. The operational determinant of the given system is
    $$L = D^3 - 3D - 2 = (D + 1)^2(D - 2),$$

    and we find that
    $$Lx = Ly = Lz = 0.$$
    Hence
    $$x = a_1e^{2t} + a_2e^{-t} + a_3te^{-t},$$
    $$y = b_1e^{2t} + b_2e^{-t} + b_3te^{-t},$$
    $$z = c_1e^{2t} + c_2e^{-t} + c_3te^{-t}.$$

    When we substitute these expressions in the three differential equations and compare coefficients of e^{2t}, we find that $a_1 = b_1 = c_1$. When we compare coefficients of te^{-t} we find that $a_3 + b_3 + c_3 = 0$. Comparison of coefficients of e^{-t} yields

$$a_2 + b_2 + c_2 = a_3 - 1 = b_3 = c_3.$$

It follows that $a_3 = 2/3$ and $b_3 = c_3 = -1/3$.
If a_2 and b_2 are chosen arbitrarily, then
$c_2 = -a_2 - b_2 - 1/3$. Hence the general solution is

$$x = a_1 e^{2t} + a_2 e^{-t} + (2/3) t e^{-t},$$
$$y = a_1 e^{2t} + b_2 e^{-t} - (1/3) t e^{-t},$$
$$z = a_1 e^{2t} - (a_2 + b_2 + 1/3) e^{-t} - (1/3) t e^{-t}.$$

23. Subtraction of the two equations yields $x + y = e^{-2t} - e^{-3t}$.
We then verify readily that any two differentiable functions
$x(t)$ and $y(t)$ satisfying this condition will constitute a
solution of the given system, which thus has infinitely many
solutions.

24. Subtraction of one equation from the other yields
$x + y = t^2 - t$. But then

$$(D + 2)x + (D + 2)y = D(x + y) + 2(x + y)$$
$$= (2t - 1) + 2(t^2 - t) = 2t^2 - 1 \neq t.$$

Thus the given system has no solution.

25. Infinitely many solutions, because any solution of the
second equation also satisfies the first equation (because
it is $D + 2$ times the second one).

26. Subtraction of the second equation from the first one gives
$x = e^{-t}$. Then substitution in the second equation yields
$D^2 y = 0$, so $y = b_1 t + b_2$. Thus there are *two* arbitrary
constants.

27. Subtraction of the second equation from the first one gives
$x + y = e^{-t}$. Then substitution in the second equation yields

$$x = D^2(x + y) = e^{-t},$$

so $y = 0$. Thus there are *no* arbitrary constants.

28. Differentiation of the difference of the two given equations yields

$$(D^2 + D)x + D^2y = -2e^{-t},$$

which contradicts the first equation. Hence the system has *no solution*.

29. Addition of the two given equations yields $D^2x = e^{-t}$,
so $x(t) = a_1t + a_2$. Then the second equation gives
$D^2y = a_1t + a_2$, so

$$y(t) = (1/6)a_1t^3 + (1/2)a_2t^2 + a_3t + a_4.$$

Thus there are *four* arbitrary constants.

30. $x(t) = a_1\exp(r_1t) + a_2\exp(r_2t)$

$y(t) = (6a_1 + 20r_1a_1)\exp(r_1t) + (6a_2 + 20r_2a_2)\exp(r_2t)$

where $r_1 = (-9 + \sqrt{33})/40,$ $r_2 = (-9 - \sqrt{33})/40,$

$a_1 + a_2 = 50,$ $r_1a_1 + r_2a_2 = -10.$

31. $I_1(t) = 2 + e^{-5t}[-2\cos(10t/\sqrt{6}) + 4\sqrt{6}\sin(10t/\sqrt{6})]$

$I_2(t) = (20/\sqrt{6})e^{-5t}\sin(10t/\sqrt{6})$

32. The solution of the system

$$2(I_1' - I_2') + 50 I_1 = 100 \sin 60t$$
$$2(I_2' - I_1') + 25 I_2 = 0$$

with $I_1(0) = I_2(0) = 0$ is

$I_1 = ($ $120e^{-25t/3} - 120 \cos 60t + 1778 \sin 60t)/1321$

$I_2 = (- 240e^{-25t/3} + 240 \cos 60t + 1728 \sin 60t)/1321$

(Yes, one cofficient of sin 60t is 1778 and the other is 1728.)

33. See the solution to Problem 30 in Section 5.1. The solution of the system

$$I_1' = -20(I_1 - I_2), \quad I_2' = 40(I_1 - I_2)$$

with $I_1(0) = 2$ and $I_2(0) = 0$ is given by

$$I_1(t) = 2(2 + e^{-60t})/3, \quad I_2(t) = 4(1 - e^{-60t})/3.$$

34. $3x_1/100 = 1 + e^{-3t/20}[2 \cos(t/20)\sqrt{3}]$

$3x_2/100 = 1 + e^{-3t/20}[- \cos(t/20)\sqrt{3} + \sqrt{3} \sin(t/20)\sqrt{3}]$

$3x_3/100 = 1 + e^{-3t/20}[- \cos(t/20)\sqrt{3} - \sqrt{3} \sin(t/20)\sqrt{3}]$

35. The two given equations yield

$$mx^{(3)} = qBy'' = -q^2B^2x'/m,$$

so $x^{(3)} + \omega^2x' = 0$. The general solution is

$$x(t) = A \cos \omega t + B \sin \omega t + C.$$

Now $x'(0) = 0$ implies $B = 0$, and then $x(0) = r_0$ gives $A + C = r_0$. Next,

$$\omega y' = x'' = -A\omega^2\cos \omega t,$$

so $y'(0) = - \omega r_0$ implies $A = r_0$, hence $C = 0$. It now follows readily that the trajectory is the circle

$$x(t) = r_0\cos \omega t, \quad y(t) = - r_0\sin \omega t.$$

36. With $\omega = qB/m$ our differential equations are

$$x'' = \omega y' + qE/m, \qquad y'' = -\omega x'.$$

Elimination gives

$$x^{(4)} + \omega^2 x'' = y^{(4)} + \omega^2 y'' = 0,$$

so

$$x = a_1 + b_1 t + c_1 \cos \omega t + d_1 \sin \omega t,$$
$$y = a_2 + b_2 t + c_2 \cos \omega t + d_2 \sin \omega t.$$

The initial conditions $x(0) = x'(0) = 0 = y(0) = y'(0)$ yield

$$c_1 = -a_1, \quad b_1 = -\omega d_1, \quad c_2 = -a_2, \quad b_2 = -\omega d_2.$$

Then substitution in $y'' = -\omega x'$ yields

$$d_1 = b_1 = a_2 = c_2 = 0 \quad \text{and} \quad d_2 = a_1.$$

Finally substitution in $x'' = \omega y' + qE/m$ yields $a = a_1 = E/\omega B$, so the solution is

$$x = a(1 - \cos \omega t), \qquad y = -a(\omega t - \sin \omega t).$$

37. (a) $x = a_1 \cos 5t + a_2 \sin 5t + b_1 \cos 5t\sqrt{3} + b_2 \sin 5t\sqrt{3}$,
$y = 2a_1 \cos 5t + 2a_2 \sin 5t - 2b_1 \cos 5t\sqrt{3} - 2b_2 \sin 5t\sqrt{3}$

(b) In the natural mode with frequency $\omega_1 = 5$ the masses move in the same direction, while in the natural mode with frequency $\omega_2 = 5\sqrt{3}$ they move in opposite directions. In each case the amplitude of the motion of m_2 is twice that of m_1.

39. $x(t) = a_1 \cos t + a_2 \sin t + b_1 \cos 3t + b_2 \sin 3t$
$y(t) = a_1 \cos t + a_2 \sin t - b_1 \cos 3t - b_2 \sin 3t$

40. $x = (a_1 \cos t + a_2 \sin t) + (b_1 \cos 2t + b_2 \sin 2t)$
$y = (a_1 \cos t + a_2 \sin t) - (b_1 \cos 2t + b_2 \sin 2t)/2$

In the natural mode with frequency $\omega_1 = 1$ the two masses move in the same direction with equal amplitudes of oscillation. In the natural mode with frequency $\omega_2 = 2$ the two masses move in opposite directions with the amplitude of m_2 being half that of m_1.

41. $x = a_1\cos t + a_2\sin t + b_1\cos t\sqrt{5} + b_2\sin t\sqrt{5}$
     $y = a_1\cos t + a_2\sin t - b_1\cos t\sqrt{5} - b_2\sin t\sqrt{5}$

42. $x = (a_1\cos t\sqrt{2} + a_2\sin t\sqrt{2}) + (b_1\cos 2t + b_2\sin 2t)$
     $y = (a_1\cos t\sqrt{2} + a_2\sin t\sqrt{2}) - (b_1\cos 2t + b_2\sin 2t)$

In the natural mode with frequency $\omega_1 = \sqrt{2}$ the two masses move in the same direction; in the natural mode with frequency $\omega_2 = 2$ they move in opposite directions. In each natural mode their amplitudes of oscillation are equal.

43. $x = a_1\cos t\sqrt{2} + a_2\sin t\sqrt{2} + b_1\cos t\sqrt{8} + b_2\sin t\sqrt{8}$
     $y = a_1\cos t\sqrt{2} + a_2\sin t\sqrt{2} - (b_1\cos t\sqrt{8} + b_2\sin t\sqrt{8})/2$

44. (b) The operational determinant is

$$(D^2 + 2)[(D^2 + 2)^2 - 1] + [-(D^2 + 2)]$$
$$= (D^2 + 2)[(D^2 + 2)^2 - 2],$$

and the equation

$$(r^2 + 2)[(r^2 + 2)^2 - 2] = 0$$

has roots $\pm i\sqrt{2}$ and $\pm i(2 \pm \sqrt{2})^{1/2}$.

Chapter 5

LINEAR SYSTEMS AND MATRICES

The first half-dozen pages of this section are devoted to a review of matrix notation and terminology. With students who've had some prior exposure to matrices and determinants, this review material can be skimmed rapidly. In this event serious study of the section can begin with the subsections on matrix-valued functions and first order linear systems. About all that's actually needed for this purpose is some acquaintance with determinants, with matrix multiplication and inverse matrices, and with the fact that a square matrix is invertible if and only if its determinant is nonzero.

1. (a) $\begin{bmatrix} 13 & -18 \\ 23 & 17 \end{bmatrix}$ (b) $\begin{bmatrix} 0 & -1 \\ 2 & 19 \end{bmatrix}$

    (c) $\begin{bmatrix} -9 & -11 \\ 47 & -9 \end{bmatrix}$ (d) $\begin{bmatrix} -10 & -37 \\ 14 & -8 \end{bmatrix}$

2. $(AB)C = A(BC) = \begin{bmatrix} -33 & -7 \\ -27 & 103 \end{bmatrix}$

    $A(B + C) = AB + AC = \begin{bmatrix} -18 & -4 \\ 68 & -8 \end{bmatrix}$

3. $AB = \begin{bmatrix} -1 & 8 \\ 46 & -1 \end{bmatrix}$; $BA = \begin{bmatrix} 11 & -12 & 14 \\ -14 & 0 & 7 \\ 0 & 8 & -13 \end{bmatrix}$

4. $Ay = \begin{bmatrix} 2t^2 - \cos t \\ 3t^2 - 4\sin t + 5\cos t \end{bmatrix}$

$$Bx = \begin{bmatrix} 2t + 3e^{-t} \\ -14t \\ 6t - 2e^{-t} \end{bmatrix}$$

The products Ax and By are not defined, because in neither case is the number of columns of the first factor equal to the number of rows of the second factor.

5. (a) $\begin{bmatrix} 21 & 2 & 1 \\ 4 & 44 & 9 \\ -27 & 34 & 45 \end{bmatrix}$ (b) $\begin{bmatrix} 9 & 21 & -13 \\ -5 & -8 & 24 \\ -25 & -19 & 26 \end{bmatrix}$

(c) $\begin{bmatrix} 0 & -6 & 1 \\ 10 & 31 & -15 \\ 16 & 58 & -23 \end{bmatrix}$ (d) $\begin{bmatrix} -10 & -8 & 5 \\ 18 & 12 & -10 \\ 11 & 22 & 6 \end{bmatrix}$

(e) $\begin{bmatrix} 3 - t & 2 & -1 \\ 0 & 4 - t & 3 \\ -5 & 2 & 7 - t \end{bmatrix}$

6. (a) $A_1 B = A_2 B = \begin{bmatrix} 5 & 10 \\ -4 & -8 \end{bmatrix}$

7. $\det(A) = \det(B) = 0$

8. $\det(AB) = \det(BA) = 144$

9. $(AB)' = \begin{bmatrix} 1 - 8t + 18t^2 & 1 + 2t - 12t^2 + 32t^3 \\ 3 + 3t^2 - 4t^3 & 8t + 3t^2 + 4t^3 \end{bmatrix}$

10. $(AB)' = A'B + AB' = \begin{bmatrix} 3e^t + 2e^{-t} - 2te^{-t} + 9t^2 \\ 3 \\ 24 + 2e^{-t} + 12t^3 \end{bmatrix}$

11. $\mathbf{x} = \begin{bmatrix} x \\ y \end{bmatrix}$, $\quad \mathbf{P}(t) = \begin{bmatrix} 0 & -3 \\ 3 & 0 \end{bmatrix}$, $\quad \mathbf{f}(t) = \begin{bmatrix} 0 \\ 0 \end{bmatrix}$

12. $\mathbf{x} = \begin{bmatrix} x \\ y \end{bmatrix}$, $\quad \mathbf{P}(t) = \begin{bmatrix} 3 & -2 \\ 2 & 1 \end{bmatrix}$, $\quad \mathbf{f}(t) = \begin{bmatrix} 0 \\ 0 \end{bmatrix}$

13. $\mathbf{P}(t) = \begin{bmatrix} 2 & 4 \\ 5 & -1 \end{bmatrix}$; $\quad \mathbf{f}(t) = \begin{bmatrix} 3e^t \\ -t^2 \end{bmatrix}$

14. $\mathbf{x} = \begin{bmatrix} x \\ y \end{bmatrix}$, $\quad \mathbf{P}(t) = \begin{bmatrix} t & -e^t \\ e^{-t} & t^2 \end{bmatrix}$, $\quad \mathbf{f}(t) = \begin{bmatrix} \cos t \\ -\sin t \end{bmatrix}$

15. $\mathbf{x} = \begin{bmatrix} x \\ y \\ z \end{bmatrix}$, $\quad \mathbf{P}(t) = \begin{bmatrix} 0 & 1 & 1 \\ 1 & 0 & 1 \\ 1 & 1 & 0 \end{bmatrix}$, $\quad \mathbf{f}(t) = \begin{bmatrix} 0 \\ 0 \\ 0 \end{bmatrix}$

16. $\mathbf{x} = \begin{bmatrix} x \\ y \\ z \end{bmatrix}$, $\quad \mathbf{P}(t) = \begin{bmatrix} 2 & -3 & 0 \\ 1 & 1 & 2 \\ 0 & 5 & -7 \end{bmatrix}$, $\quad \mathbf{f}(t) = \begin{bmatrix} 0 \\ 0 \\ 0 \end{bmatrix}$

17. $\mathbf{P}(t) = \begin{bmatrix} 3 & -4 & 1 \\ 1 & 0 & -3 \\ 0 & 6 & -7 \end{bmatrix}$; $\quad \mathbf{f}(t) = \begin{bmatrix} t \\ t^2 \\ t^3 \end{bmatrix}$

18. $\mathbf{x} = \begin{bmatrix} x \\ y \\ z \end{bmatrix}$, $\quad P(t) = \begin{bmatrix} t & -1 & e^t \\ 2 & t^2 & -1 \\ e^{-t} & 3t & t^3 \end{bmatrix}$, $\quad f(t) = \begin{bmatrix} 0 \\ 0 \\ 0 \end{bmatrix}$

19. $\mathbf{x} = \begin{bmatrix} x_1 \\ x_2 \\ x_3 \\ x_4 \end{bmatrix}$, $\quad \mathbf{P}(t) = \begin{bmatrix} 0 & 1 & 0 & 0 \\ 0 & 0 & 2 & 0 \\ 0 & 0 & 0 & 3 \\ 4 & 0 & 0 & 0 \end{bmatrix}$, $\quad \mathbf{f}(t) = \begin{bmatrix} 0 \\ 0 \\ 0 \\ 0 \end{bmatrix}$

20. $\mathbf{x} = \begin{bmatrix} x_1 \\ x_2 \\ x_3 \\ x_4 \end{bmatrix}$, $\quad \mathbf{P}(t) = \begin{bmatrix} 0 & 1 & 1 & 0 \\ 0 & 0 & 1 & 1 \\ 1 & 0 & 0 & 1 \\ 1 & 1 & 0 & 0 \end{bmatrix}$, $\quad \mathbf{f}(t) = \begin{bmatrix} 0 \\ t \\ t^2 \\ t^3 \end{bmatrix}$

21. $W(t) = e^{3t}$; $\quad \mathbf{x}(t) = \begin{bmatrix} 2c_1 e^t + c_2 e^{2t} \\ -3c_1 e^t - c_2 e^{2t} \end{bmatrix}$

22. $\quad x(t) = c_1 \begin{bmatrix} 1 \\ 3 \end{bmatrix} e^{3t} + c_2 \begin{bmatrix} 2 \\ 1 \end{bmatrix} e^{-2t}$

23. $W(t) = 4$; $\quad x(t) = \begin{bmatrix} c_1 e^{2t} + c_2 e^{-2t} \\ c_1 e^{2t} + 5c_2 e^{-2t} \end{bmatrix}$

24. $\quad x(t) = c_1 \begin{bmatrix} 1 \\ -1 \end{bmatrix} e^{3t} + c_2 \begin{bmatrix} 2 \\ -1 \end{bmatrix} e^{2t}$

25. $W(t) = 7e^{-3t}$; $\quad x(t) = \begin{bmatrix} 3c_1 e^{2t} + c_2 e^{-5t} \\ 2c_1 e^{2t} + 3c_2 e^{-5t} \end{bmatrix}$

26. $x(t) = c_1 \begin{bmatrix} 2 \\ 2 \\ 1 \end{bmatrix} e^t + c_2 \begin{bmatrix} -2 \\ 0 \\ 1 \end{bmatrix} e^{3t} + c_3 \begin{bmatrix} 2 \\ -2 \\ 1 \end{bmatrix} e^{5t}$

27. $W(t) = 3$; $\quad x(t) = \begin{bmatrix} c_1 e^{2t} + c_2 e^{-t} \\ c_1 e^{2t} + c_3 e^{-t} \\ c_1 e^{2t} - (c_2 + c_3) e^{-t} \end{bmatrix}$

28. $x(t) = c_1 \begin{bmatrix} 1 \\ 6 \\ -13 \end{bmatrix} + c_2 \begin{bmatrix} 2 \\ 3 \\ -2 \end{bmatrix} e^{3t} + c_3 \begin{bmatrix} -1 \\ 2 \\ 1 \end{bmatrix} e^{-4t}$

29. $W(t) = e^{2t}$; $\quad \mathbf{x}(t) = \begin{bmatrix} 3c_1 e^{-2t} + c_2 e^t + c_3 e^{3t} \\ -2c_1 e^{-2t} - c_2 e^t - c_3 e^{3t} \\ 2c_1 e^{-2t} + c_2 e^t \end{bmatrix}$

30. $W(t) = 1$; $\quad \mathbf{x}(t) = \begin{bmatrix} c_1 e^{-t} + c_4 e^t \\ c_3 e^t \\ c_2 e^{-t} + 3c_4 e^t \\ c_1 e^{-t} - 2c_3 e^t \end{bmatrix}$

31. $\bar{x} = 2\bar{x}_1 - \bar{x}_2$

32. $\bar{x} = 7\bar{x}_1 - 2\bar{x}_2$

33. $\bar{x} = 3\bar{x}_1 + 4\bar{x}_2$

34. $\bar{x} = 3\bar{x}_1 - 2\bar{x}_2$

35. $\bar{x} = \bar{x}_1 + 2\bar{x}_2 + \bar{x}_3$

36. $\bar{x} = 7\bar{x}_1 + 3\bar{x}_2 + 5\bar{x}_3$

37. $\mathbf{x} = 3x_1^- - 3x_2^- - 5x_3^-$

38. $\bar{x} = -2\bar{x}_1 + 15\bar{x}_2 - 4\bar{x}_3$

39. $\bar{x} = 3\bar{x}_1 + 7\bar{x}_2 + \bar{x}_3 - 2x_4^-$

40. $\bar{x} = 13\bar{x}_1 + 41\bar{x}_2 + 3\bar{x}_3 - 12x_4^-$

41. (a) $x_2 = tx_1$, so neither is a constant multiple of the other.

(b) $W(x_1, x_2) = 0$, whereas Theorem 2 would imply that $W \neq 0$ if x_1 and x_2 were independent solutions of a system of the indicated form.

42. If $x_{12}(t) = c\, x_{11}(t)$ and $x_{22}(t) = c\, x_{21}(t)$

then

$$W(t) = x_{11}(t)x_{22}(t) - x_{12}(t)x_{21}(t)$$

$$= c\, x_{11}(t)x_{21}(t) - c\, x_{11}(t)x_{21}(t) = 0.$$

SECTION 5.4

THE EIGENVALUE METHOD FOR HOMOGENEOUS LINEAR SYSTEMS

1. Eigenvalues $\lambda_1 = -1$ and $\lambda_2 = 3$

   Eigenvectors $v_1 = [1 \quad -1]^T$ and $v_2 = [1 \quad 1]^T$

$$x_1 = c_1e^{-t} + c_2e^{3t}$$
$$x_2 = -c_1e^{-t} + c_2e^{3t}$$

2. Eigenvalues $\lambda_1 = -1$ and $\lambda_2 = 4$

   Eigenvectors $v_1 = [1 \quad -1]^T$ and $v_2 = [3 \quad 2]^T$

$$x_1 = c_1e^{-t} + 3c_2e^{4t}$$
$$x_2 = -c_1e^{-t} + 2c_2e^{4t}$$

3. Eigenvalues $\lambda_1 = -1$ and $\lambda_2 = 6$

   Eigenvectors $v_1 = [1 \quad -1]^T$ and $v_2 = [4 \quad 3]^T$

   General solution
   $$x_1 = c_1e^{-t} + 4c_2e^{6t}$$
   $$x_2 = -c_1e^{-t} + 3c_2e^{6t}$$

   The equations
   $$x_1(0) = c_1 + 4c_2 = 1$$
   $$x_2(0) = -c_1 + 3c_2 = 1$$

   yield $c_1 = -1/7$ and $c_2 = 2/7$, so the desired particular solution is given by

   $$x_1 = (-e^{-t} + 8e^{6t})/7$$
   $$x_2 = (e^{-t} + 6e^{6t})/7.$$

4. Eigenvalues $\lambda_1 = -2$ and $\lambda_2 = 5$

   Eigenvectors $v_1 = [1 \quad -6]^T$ and $v_2 = [1 \quad 1]^T$

$$x_1 = \quad c_1 e^{-2t} + c_2 e^{5t}$$
$$x_2 = - 6c_1 e^{-2t} + c_2 e^{5t}$$

5. Eigenvalues $\lambda_1 = - 1$ and $\lambda_2 = 5$

    Eigenvectors $v_1 = [1 \quad 1]^T$ and $v_2 = [7 \quad 1]^T$

$$x_1 = c_1 e^{-t} + 7c_2 e^{5t}$$
$$x_2 = c_1 e^{-t} + \quad c_2 e^{5t}$$

6. Eigenvalues $\lambda_1 = 3$ and $\lambda_2 = 4$

    Eigenvectors $v_1 = [5 \quad -6]^T$ and $v_2 = [1 \quad -1]^T$

$$x_1 = \quad 5c_1 e^{3t} + c_2 e^{4t}$$
$$x_2 = - 6c_1 e^{3t} - c_2 e^{4t}$$

The initial conditions yield $c_1 = -1$ and $c_2 = 6$, so

$$x_1 = - 5e^{3t} + 6e^{4t}, \quad x_2 = 6e^{3t} - 6e^{4t}.$$

7. Eigenvalues $\lambda_1 = 1$ and $\lambda_2 = - 9$

    Eigenvectors $v_1 = [1 \quad 1]^T$ and $v_2 = [2 \quad -3]^T$

$$x_1 = c_1 e^{t} + 2c_2 e^{-9t}$$
$$x_2 = c_1 e^{t} - 3c_2 e^{-9t}$$

8. Characteristic equation $\lambda^2 + 4 = 0$

    Eigenvalue $\lambda = 2i$, eigenvector $v = [5 \quad 1-2i]^T$

$$x_1 = 5c_1\cos 2t + 5c_2\sin 2t$$

$$x_2 = c_1(\cos 2t + 2\sin 2t) + c_2(\sin 2t - 2\cos 2t)$$
$$= (c_1 - 2c_2)\cos 2t + (2c_1 + c_2)\sin 2t$$

9. Characteristic equation $\lambda^2 + 16 = 0$

Eigenvalue $\lambda = 4i$, eigenvector $v = [5 \quad 2-4i]^T$

The real and imaginary parts of

$$x = [5 \quad 2-4i]^T(\cos 4t + i\sin 4t)$$

yield the general solution

$$x_1 = 5c_1\cos 4t + 5c_2\sin 4t$$
$$x_2 = c_1(2\cos 4t + 4\sin 4t) + c_2(2\sin 4t - 4\cos 4t).$$

The initial conditions $x_1(0) = 2$ and $x_2(0) = 3$ give
$c_1 = 2/5$ and $c_2 = -11/20$, so the desired particular
solution is

$$x_1 = 2\cos 4t - (11/4)\sin 4t$$
$$x_2 = 3\cos 4t + (1/2)\sin 4t.$$

10. Characteristic equation $\lambda^2 + 9 = 0$

Eigenvalue $\lambda = 3i$, eigenvector $v = [-2 \quad 3+2i]^T$

$$x_1 = -2c_1\cos 3t - 2c_2\sin 3t$$

$x_2 = c_1(3 \cos 3t - 3 \sin 3t) + c_2(3 \cos 3t + 3 \sin 3t)$

$\quad = (3c_1 + 3c_2)\cos 3t + (3c_2 - 3c_1)\sin 3t$

11. Characteristic equation $(\lambda - 1)^2 + 4 = 0$

Eigenvalue $\lambda = 1 - 2i$, eigenvector $v = [1 \quad i]^T$

The real and imaginary parts of

$\quad x = [1 \quad i]^T e^t(\cos 2t - i \sin 2t)$

$\quad\quad = e^t[\cos 2t \quad \sin 2t]^T + ie^t[-\sin 2t \quad \cos 2t]^T$

yield the general solution

$\quad\quad x_1 = e^t(c_1\cos 2t - c_2\sin 2t)$

$\quad\quad x_2 = e^t(c_1\sin 2t + c_2\cos 2t)$.

The particular solution with $x_1(0) = 0$ and $x_2(0) = 4$
is obtained with $c_1 = 0$ and $c_2 = 4$, so

$\quad\quad x_1 = -4e^t\sin 2t, \quad x_2 = 4e^t\cos 2t.$

12. Characteristic equation $\lambda^2 - 4\lambda + 8 = 0$

Eigenvalue $\lambda = 2 + 2i$, eigenvector $v = [-5 \quad 1+2i]^T$

$x_1 = e^{2t}(- 5c_1\cos 2t - 5c_2\sin 2t)$

$x_2 = e^{2t}[c_1(\cos 2t - 2 \sin 2t) + c_2(2 \cos 2t + \sin 2t)]$

$\quad = e^{2t}[(c_1 + 2c_2)\cos 2t + (- 2c_1 + c_2)\sin 2t]$

13. Characteristic equation $\lambda^2 - 4\lambda + 13 = 0$

Eigenvalue $\lambda = 2 - 3i$, eigenvector $v = \begin{bmatrix} 3 & 1+i \end{bmatrix}^T$

The real and imaginary parts of

$$x = e^{2t}\begin{bmatrix} 3 & 1+i \end{bmatrix}^T(\cos 3t - i \sin 3t)$$

yield the general solution

$$x_1 = 3e^{2t}(c_1\cos 3t - c_2\sin 3t)$$
$$x_2 = e^{2t}[(c_1 + c_2)\cos 3t + (c_1 - c_2)\sin 3t)].$$

14. Characteristic equation $(3 - \lambda)^2 + 16 = 0$

Eigenvalue $\lambda = 3 + 4i$, eigenvector $v = \begin{bmatrix} 1 & -i \end{bmatrix}^T$

$$x_1 = e^{3t}(c_1\cos 4t + c_2\sin 4t)$$
$$x_2 = e^{3t}(c_1\sin 4t - c_2\cos 4t)$$

15. Characteristic equation $\lambda^2 - 10\lambda + 41 = 0$

Eigenvalue $\lambda = 5 - 4i$, eigenvector $v = \begin{bmatrix} 5 & 2+4i \end{bmatrix}^T$

$$x = 5e^{5t}\begin{bmatrix} 5 & 2+4i \end{bmatrix}^T(\cos 4t - i \sin 4t)$$

$$x_1 = 5e^{5t}(c_1\cos 4t - c_2\sin 4t)$$
$$x_2 = e^{5t}[(2c_1 + 4c_2)\cos 4t + (4c_1 - 2c_2)\sin 4t)].$$

16. Characteristic equation $\lambda^2 + 110\lambda + 1000 = 0$

Eigenvalues $\lambda_1 = -10$ and $\lambda_2 = -100$

Eigenvectors $v_1 = [1 \ 2]^T$ and $v_2 = [2 \ -5]^T$

$$x_1 = c_1 e^{-10t} + 2c_2 e^{-100t}$$
$$x_2 = 2c_1 e^{-10t} - 5c_2 e^{-100t}$$

17. Eigenvalues $\lambda_1 = 9$, $\lambda_2 = 6$, $\lambda_3 = 0$

Eigenvectors $v_1[1 \ \ 1 \ \ 1]^T$, $v_2 = [1 \ \ -2 \ \ 1]^T$,

$v_3[1 \ \ 0 \ \ -1]^T$

$$x_1 = c_1 e^{9t} + c_2 e^{6t} + c_3$$
$$x_2 = c_1 e^{9t} - 2c_2 e^{6t}$$
$$x_3 = c_1 e^{9t} + c_2 e^{6t} - c_3$$

18. Eigenvalues $\lambda_1 = 9$, $\lambda_2 = 6$, $\lambda_3 = 0$

Eigenvectors $v_1[1 \ \ 2 \ \ 2]^T$, $v_2 = [0 \ \ 1 \ \ -1]^T$,

$v_3[4 \ \ -1 \ \ -1]^T$

$$x_1 = c_1 e^{9t} \qquad\qquad + 4c_3$$
$$x_2 = 2c_1 e^{9t} + c_2 e^{6t} - c_3$$
$$x_3 = 2c_1 e^{9t} - c_2 e^{6t} - c_3$$

19. Eigenvalues $\lambda_1 = 6$, $\lambda_2 = 3$, $\lambda_3 = 3$

Eigenvectors $v_1[1 \ \ 1 \ \ 1]^T$, $v_2 = [1 \ \ -2 \ \ 1]^T$,

$v_3[1 \ \ 0 \ \ -1]^T$

$$x_1 = c_1 e^{6t} + c_2 e^{3t} + c_3 e^{3t}$$

$$x_2 = c_1e^{6t} - 2c_2e^{3t}$$

$$x_3 = c_1e^{6t} + c_2e^{3t} - c_3e^{3t}$$

20. Eigenvalues $\lambda_1 = 9,$ $\lambda_2 = 6,$ $\lambda_3 = 2$

Eigenvectors $v_1[1 \quad 1 \quad 1]^T,$ $v_2 = [1 \quad -2 \quad 1]^T,$

$v_3[1 \quad 0 \quad -1]^T$

$$x_1 = c_1e^{9t} + c_2e^{6t} + c_3e^{2t}$$

$$x_2 = c_1e^{9t} - 2c_2e^{6t}$$

$$x_3 = c_1e^{9t} + c_2e^{6t} - c_3e^{2t}$$

21. Eigenvalues $\lambda_1 = 0,$ $\lambda_2 = 1,$ $\lambda_3 = -1$

Eigenvectors $v_1[6 \quad 2 \quad 5]^T,$ $v_2 = [3 \quad 1 \quad 2]^T,$

$v_3[2 \quad 1 \quad 2]^T$

$$x_1 = 6c_1 + 3c_2e^t + 2c_3e^t$$

$$x_2 = 2c_1 + c_2e^t \qquad c_3e^t$$

$$x_3 = 5c_1 + 2c_2e^t + 2c_3e^t$$

22. Characteristic equation $(3 - \lambda)(\lambda^2 + \lambda - 2) = 0$

Distinct eigenvalues $\lambda_1 = -2,$ $\lambda_2 = 1,$ $\lambda_3 = 3$

Eigenvectors $v_1 = [0 \quad 1 \quad -1]^T,$ $v_2 = [1 \quad -1 \quad 0]^T,$

$v_3 = [1 \quad -1 \quad 1]^T$

$$x_1 = \qquad\qquad c_2e^t + c_3e^{3t}$$

$$x_2 = c_1e^{-2t} - c_2e^t - c_3e^{3t}$$

$$x_3 = -c_1e^{-2t} \qquad + c_3e^{3t}$$

23. Eigenvalues $\lambda_1 = 2$, $\lambda_2 = -2$, $\lambda_3 = 3$

Eigenvectors $v_1 = [1 \quad -1 \quad 0]^T$, $v_2 = [0 \quad 1 \quad -1]^T$,

$$v_3 = [1 \quad -1 \quad 1]^T$$

$$x_1 = c_1 e^{2t} \qquad\qquad + c_3 e^{3t}$$
$$x_2 = -c_1 e^{2t} + c_2 e^{-2t} - c_3 e^{3t}$$
$$x_3 = \qquad\qquad - c_2 e^{-2t} + c_3 e^{3t}$$

24. Characteristic equation $4(1 - \lambda)(\lambda^2 + 4) = 0$

Eigenvalues $\lambda = 1$ and $\lambda = \pm 2i$

Eigenvector $[1 \quad -1 \quad 0]^T$ associated with $\lambda = 1$

To find an eigenvector $v = [a \quad b \quad c]^T$ associated with $\lambda = 2i$ we must find a nontrivial solution of the equations

$$(2 - 2i)a + \qquad\qquad b - \qquad\qquad c = 0,$$
$$-4a + (-3 - 2i)b - \qquad\qquad c = 0,$$
$$4a + \qquad\qquad 4b + (2 - 2i)c = 0.$$

Subtraction of the first two equations yields

$$(6 - 2i)a + (4 + 2i)b = 0,$$

so we take $a = 2 + i$ and $b = -3 + i$. Then the first equation gives $c = 3 - i$. Thus $v = [2+i \quad -3+i \quad 3-i]^T$. Finally

$$(2 + i)e^{2it} = (2 \cos 2t - \sin 2t) + i(\cos 2t + 2 \sin 2t),$$
$$(3 - i)e^{2it} = (3 \cos 2t + \sin 2t) + i(3 \sin 2t - \cos 2t),$$

so the solution is

$$x_1 = c_1 e^t + c_2(2 \cos 2t - \sin 2t) + c_3(\cos 2t + 2 \sin 2t)$$

$$x_2 = - c_1 e^t - c_2(3 \cos 2t + \sin 2t) + c_3(\cos 2t - 3 \sin 2t)$$

$$x_3 = c_2(3 \cos 2t + \sin 2t) + c_3(3 \sin 2t - \cos 2t)$$

25. Eigenvalues $\lambda = 0, \quad 2 \pm 3i$

$$x_1 = c_1 + e^{2t}[(-c_2 + c_3)\cos 3t + (c_2 + c_3)\sin 3t]$$

$$x_2 = - c_1 + 2e^{2t}(c_2\cos 3t - c_3\sin 3t)$$

$$x_3 = 2e^{2t}(-c_2\cos 3t + c_3\sin 3t)$$

26. Characteristic equation $(3 - \lambda)[(\lambda + 1)^2 + 1] = 0$

Eigenvalues $\lambda = 3$ and $\lambda = -1 \pm i$

Eigenvector $v_0 = [4 \quad 9 \quad 0]$ associated with $\lambda_0 = 3$.

To find an eigenvector $v = [a \quad b \quad c]^T$ associated with the

complex eigenvalue $\lambda = -1 + i$ we must find a nontrivial

solution of the equations

$$(4 - i)a + c = 0$$

$$9a - ib + 2c = 0$$

$$-9a + 4b - ic = 0$$

The choice $a = 1$ in the first equation gives $c = -4 + i$.

Then the third equation yields $b = 2 - i$. Thus the complex

eigenvector associated with $\lambda = -1 + i$ is

$$v = [1 \quad 2-i \quad -4+i]^T.$$

Finding the real and imaginary parts $x_1(t)$ and $x_2(t)$ of

$$x(t) = ve^{(-1+i)t} = ve^{-t}(\cos t + i \sin t)$$

and assembling the general solution $c_0 x_0 + c_1 x_1 + c_2 x_2$, we get the scalar equations

$$x_1(t) = 4c_0 e^{et} + e^{-t}[c_1 \cos t + c_2 \sin t]$$
$$x_2(t) = 9c_0 e^{3t} + e^{-t}[(2c_1 - c_2)\cos t + (c_1 + 2c_2)\sin t]$$
$$x_3(t) = \qquad\qquad e^{-t}[(-4c_1 + c_2)\cos t + (-c_1 - 4c_2)\sin t]$$

Finally, the given initial conditions yield the values $c_0 = 1$, $c_1 = -4$, $c_2 = 1$, so the desired particular solution is

$$x_1(t) = 4e^{3t} - e^{-t}(4 \cos t - \sin t)$$
$$x_2(t) = 9e^{3t} - e^{-t}(9 \cos t + 2 \sin t)$$
$$x_3(t) = 17e^{-t}\cos t.$$

27. The matrix

$$A = \begin{bmatrix} -0.2 & 0 \\ 0.2 & -0.4 \end{bmatrix}$$

has eigenvalues $\lambda_1 = -0.2$ and $\lambda_2 = -0.4$ with eigenvectors $v_1 = [1 \quad 1]^T$ and $v_2 = [0 \quad 1]^T$.

$$x_1(t) = 15e^{-0.2t}, \quad x_2(t) = 15e^{-0.2t} - 15e^{-0.4t}$$

The maximum value of x_2, when $x_2'(t) = 0$, is 3.75 lb.

28. The matrix

$$A = \begin{bmatrix} -0.4 & 0 \\ 0.4 & -0.25 \end{bmatrix}$$

has eigenvalues $\lambda_1 = -0.4$ and $\lambda_2 = -0.25$
with eigenvectors $v_1 = [3 \quad -8]^T$ and $v_2 = [0 \quad 1]^T$.

$$x_1(t) = 15e^{-0.4t}, \quad x_2(t) = -40e^{-0.4t} + 40e^{-0.25t}$$

The maximum value of x_2, when $x_2'(t) = 0$, is 6.85 lb.

29. The matrix

$$A = \begin{bmatrix} -0.2 & 0.4 \\ 0.2 & -0.4 \end{bmatrix}$$

has eigenvalues $\lambda_1 = 0$ and $\lambda_2 = -0.6$
with eigenvectors $v_1 = [2 \quad 1]^T$ and $v_2 = [1 \quad -1]^T$.

$$x_1(t) = 10 + 5e^{-0.6t}, \quad x_2(t) = 5 - 5e^{-0.6t}$$

30. The matrix

$$A = \begin{bmatrix} -0.4 & 0.25 \\ 0.4 & -0.25 \end{bmatrix}$$

has eigenvalues $\lambda_1 = 0$ and $\lambda_2 = -0.65$
with eigenvectors $v_1 = [5 \quad 8]^T$ and $v_2 = [1 \quad -1]^T$.

$$x_1(t) = (\ 75 + 120e^{-0.65t})/13$$
$$x_2(t) = (120 - 120e^{-0.65t})/13$$

31. The matrix

$$A = \begin{bmatrix} -0.5 & 0 & 0 \\ 0.5 & -0.2 & 0 \\ 0 & 0.2 & -0.25 \end{bmatrix}$$

has eigenvalues $\lambda_1 = -0.5$, $\lambda_2 = -0.2$, and $\lambda_3 = -0.25$
with eigenvectors $v_1 = [3 \quad -5 \quad 4]^T$, $v_2 = [0 \quad 1 \quad 4]^T$, and

$$v_3 = [0 \quad 0 \quad 1]^T.$$

$$x_1(t) = 3c_1 e^{-0.5t}$$
$$x_2(t) = -5c_1 e^{-0.5t} + c_2 e^{-0.2t}$$
$$x_3(t) = 4c_1 e^{-0.5} + 4c_2 e^{-0.2t} + c_3 e^{-0.25t}$$

32. The matrix

$$A = \begin{bmatrix} -0.2 & 0 & 0 \\ 0.2 & -0.4 & 0 \\ 0 & 0.4 & -0.5 \end{bmatrix}$$

has eigenvalues $\lambda_1 = -0.2$, $\lambda_2 = -0.4$, and $\lambda_3 = -0.5$ with eigenvectors $v_1 = [3 \quad 3 \quad 4]^T$, $v_2 = [0 \quad 1 \quad 4]^T$, and $v_3 = [0 \quad 0 \quad 1]^T$.

$$x_1(t) = 3c_1 e^{-0.2t}$$
$$x_2(t) = 3c_1 e^{-0.2t} + c_2 e^{-0.4t}$$
$$x_3(t) = 4c_1 e^{-0.2t} + 4c_2 e^{-0.4t} + c_3 e^{-0.5t}$$

33. The matrix

$$A = \begin{bmatrix} -0.5 & 0 & 0.5 \\ 0.5 & -0.2 & 0 \\ 0 & 0.2 & -0.5 \end{bmatrix}$$

has real eigenvalue $\lambda = 0$ with eigenvector $v = [2 \quad 5 \quad 2]^T$, and complex eigenvalue $\lambda = -0.6 - 0.3i$ with complex eigenvector $[5 \quad -1-3i \quad -4+3i]^T$.

$$x_1(t) = 2c_0 + e^{-0.6t}(5c_1 \cos 0.3t - 5c_2 \sin 0.3t)$$
$$x_2(t) = 5c_0 + e^{-0.6t}[(-c_1 - 3c_2)\cos 0.3t$$
$$+ (-3c_1 + c_2)\sin 0.3t]$$
$$x_3(t) = 2c_0 + e^{-0.6t}[-4c_1 + 3c_2)\cos 0.3t$$
$$+ (3c_1 + 4c_2)\sin 0.3t]$$

34. Characteristic equation $(\lambda - 1)(\lambda - 2)(\lambda - 3)(\lambda - 4) = 0$
Eigenvalues and associated eigenvectors:

$$\lambda = 1, \quad v = [1 \quad -2 \quad 3 \quad -4]^T$$
$$\lambda = 2, \quad v = [0 \quad 1 \quad -3 \quad 6]^T$$
$$\lambda = 3, \quad v = [0 \quad 0 \quad 1 \quad -4]^T$$
$$\lambda = 4, \quad v = [0 \quad 0 \quad 0 \quad 1]^T$$

$$x_1(t) = c_1 e^t$$
$$x_2(t) = -2c_1 e^t + c_2 e^{2t}$$
$$x_3(t) = 3c_1 e^t - 3c_2 e^{2t} - c_3 e^{3t}$$
$$x_4(t) = -4c_1 e^t + 6c_2 e^{2t} - 4c_3 e^{3t} + c_4 e^{4t}$$

35. Characteristic equation $(\lambda^2 - 1)(\lambda^2 - 4) = 0$
Eigenvalues and associated eigenvectors:

$$\lambda = 1, \quad v = [3 \quad -2 \quad 4 \quad 1]^T$$
$$\lambda = -1, \quad v = [0 \quad 0 \quad 1 \quad 0]^T$$
$$\lambda = 2, \quad v = [0 \quad 1 \quad 0 \quad 0]^T$$
$$\lambda = -2, \quad v = [1 \quad -1 \quad 0 \quad 0]^T$$

$$x_1(t) = 3c_1 e^t \qquad\qquad\qquad + c_4 e^{-2t}$$
$$x_2(t) = -2c_1 e^t \qquad\qquad + c_3 e^{2t} - c_4 e^{-2t}$$
$$x_3(t) = 4c_1 e^t + c_2 e^{-t}$$
$$x_4(t) = c_1 e^t$$

36. Characteristic equation $(\lambda^2 - 4)(\lambda^2 - 25) = 0$
Eigenvalues and associated eigenvectors:

$$\lambda = 2, \quad v = [1 \quad -3 \quad 0 \quad 0]^T$$
$$\lambda = -2, \quad v = [0 \quad 3 \quad 0 \quad -1]^T$$
$$\lambda = 5, \quad v = [0 \quad 0 \quad 1 \quad -3]^T$$
$$\lambda = -5, \quad v = [0 \quad 1 \quad 0 \quad 0]^T$$

$$x_1(t) = c_1 e^{2t}$$
$$x_2(t) = -3c_1 e^{2t} + 3c_2 e^{-2t} \qquad\qquad - c_4 e^{-5t}$$
$$x_3(t) = \qquad\qquad\qquad\qquad c_3 e^{5t}$$
$$x_4(t) = \qquad - c_2 e^{-2t} - 3c_3 e^{5t}$$

37. The eigenvectors associated with the respective eigenvalues

$$\lambda_1 = -3, \quad \lambda_2 = -6, \quad \lambda_3 = 10, \quad \text{and} \quad \lambda_4 = 15 \quad \text{are}$$

$$\mathbf{v}_1 = [\ 1 \quad 0 \quad 0 \quad -1\ ]^T$$
$$\mathbf{v}_2 = [\ 0 \quad 1 \quad -1 \quad 0\ ]^T$$
$$\mathbf{v}_3 = [-2 \quad 1 \quad 1 \quad -2\ ]^T$$
$$\mathbf{v}_4 = [\ 1 \quad 2 \quad 2 \quad 1\ ]^T.$$

Hence the general solution is

$$
\begin{aligned}
x_1 &= c_1 e^{-3t} && - 2c_3 e^{10t} + c_4 e^{15t}\\
x_2 &= && c_2 e^{-6t} + c_3 e^{10t} + 2c_4 e^{15t}\\
x_3 &= && - c_2 e^{-6t} + c_3 e^{10t} + 2c_4 e^{15t}\\
x_4 &= - c_1 e^{-3t} && - 2c_3 e^{10t} + c_4 e^{15t}.
\end{aligned}
$$

The given initial conditions are satisfied by choosing $c_1 = c_2 = 0$, $c_3 = -1$, and $c_4 = 1$, so the desired particular solution is given by

$$x_1 = 2e^{10t} + e^{15t} = x_4$$
$$x_2 = - e^{10t} + 2e^{15t} = x_3.$$

SECTION 5.5

SECOND ORDER SYSTEMS
AND MECHANICAL APPLICATIONS

This section uses the eigenvalue method to exhibit realistic applications of linear systems. If a computer system like MATLAB is available, a system of more than three railway cars, or a multi-story building with four or more floors, can be investigated. However, the problems in the text are intended for manual solution.

Problems 1-6 involve the system

$$
\begin{aligned}
m_1 x_1'' &= -(k_1 + k_2) x_1 + k_2 x_2,\\
m_2 x_2'' &= k_2 x_1 - (k_2 + k_3) x_2
\end{aligned}
$$

with various values of m_1, m_2 and k_1, k_2, k_3.

1. $x_1 = a_1 + a_2t + b_1\cos 2t + b_2\sin 2t$
    $x_2 = a_1 + a_2t - b_1\cos 2t - b_2\sin 2t$

    The natural frequencies are $\omega_1 = 0$ and $\omega_2 = 2$. In the degenerate natural mode with "frequency" $\omega_1 = 0$ the two masses move by translation without oscillating. At frequency $\omega_2 = 2$ they oscillate in opposite directions with equal amplitudes.

2. $x_1 = a_1\cos t + a_2\sin t + b_1\cos 3t + b_2\sin 3t$
    $x_2 = a_1\cos t + a_2\sin t - b_1\cos 3t - b_2\sin 3t$

    The natural frequencies are $\omega_1 = 1$ and $\omega_2 = 3$. In the natural mode with frequency ω_1, the two masses m_1 and m_2 move in the same direction with equal amplitudes of oscillation. At frequency ω_2 they move in opposite directions with equal amplitudes.

3. $x_1 = a_1\cos t + a_2\sin t - 2b_1\cos 2t - 2b_2\sin 2t$
    $x_2 = a_1\cos t + a_2\sin t + b_1\cos 2t + b_2\sin 2t$

    The natural frequencies are $\omega_1 = 1$ and $\omega_2 = 2$. In the natural mode with frequency ω_1, the two masses m_1 and m_2 move in the same direction with equal amplitudes of oscillation. In the natural mode with frequency ω_2 they move in opposite directions with the amplitude of oscillation of m_1 twice that of m_2.

4. $x_1 = a_1\cos t + a_2\sin t + b_1\cos t\sqrt{5} + b_2\sin t\sqrt{5}$
    $x_2 = a_1\cos t + a_2\sin t - b_1\cos t\sqrt{5} - b_2\sin t\sqrt{5}$

    The natural modes of oscillation are the same as in Problem 6, except with $\omega_1 = 1$ and $\omega_2 = \sqrt{5}$.

5. $x_1 = a_1\cos t\sqrt{2} + a_2\sin t\sqrt{2} + b_1\cos 2t + b_2\sin 2t$
    $x_2 = a_1\cos t\sqrt{2} + a_2\sin t\sqrt{2} - b_1\cos 2t - b_2\sin 2t$

    The natural modes of oscillation are the same as in Problem 6, except with $\omega_1 = \sqrt{2}$ and $\omega_2 = 2$.

6. $x_1 = a_1\cos t\sqrt{2} + a_2\sin t\sqrt{2} + \quad b_1\cos 2t\sqrt{2} + \quad b_2\sin 2t\sqrt{2}$
    $x_2 = a_1\cos t\sqrt{2} + a_2\sin t\sqrt{2} - (b_1/2)\cos 2t\sqrt{2} - (b_2/2)\sin 2t\sqrt{2}$

    At natural frequency $\omega_1 = \sqrt{2}$ the two masses move in the same direction with equal amplitudes of oscillation. At natural frequency $\omega_2 = 2\sqrt{2}$ they move in opposite directions with the amplitude of m_2 being half that of m_1.

7. The characteristic polynomial of A is

    $$\lambda^3 + 6\lambda^2 + 10\lambda + 4 = (\lambda + 2)(\lambda^2 + 4\lambda + 2)$$

    Natural frequency $\omega_1 = \sqrt{(2 - \sqrt{2})}$
          with amplitude ratios $1:\sqrt{2}:1$

    Natural frequency $\omega_2 = \sqrt{2}$
          with amplitude ratios $1:0:-1$

    Natural frequency $\omega_3 = \sqrt{(2 + \sqrt{2})}$
          with amplitude ratios $1:-\sqrt{2}:1$

8. The characteristic polynomial of A is

    $$\lambda^3 + 12\lambda^2 + 40\lambda + 32 = (\lambda + 4)(\lambda^2 + 8\lambda + 8)$$

    Natural frequency $\omega_1 = \sqrt{(4 - 2\sqrt{2})}$
          with amplitude ratios $1:\sqrt{2}:1$

    Natural frequency $\omega_2 = 2$
          with amplitude ratios $1:0:-1$

Natural frequency $\omega_3 = \sqrt{(4 + 2\sqrt{2})}$

with amplitude ratios $1:-\sqrt{2}:1$

9. $x_1 = a_1\cos t + a_2\sin t + b_1\cos 2t + b_2\sin 2t - (7/40)\cos 3t$

$x_2 = a_1\cos t + a_2\sin t$

$- (b_1\cos 2t + b_2\sin 2t)/2 + (1/40)\cos 3t$

10. $x_1(t) = c_1\cos t + c_2\sin t$

$+ c_3\cos t\sqrt{5} + c_4\sin t\sqrt{5} + (2/3)\sin 2t$

$x_2(t) = c_1\cos t + c_2\sin t$

$- c_3\cos t\sqrt{5} - c_4\sin t\sqrt{5} - (4/3)\sin 2t$

11. $x = a_1\cos t\sqrt{2} + a_2\sin t\sqrt{2} + b_1\cos 2t + b_2\sin 2t + (4/3)\cos t$

$y = a_1\cos t\sqrt{2} + a_2\sin t\sqrt{2}$

$- b_1\cos 2t - b_2\sin 2t + (5/3)\cos t$

12.   The equations of motion of the given system are

$$x_1'' = - 50x_1 + 10(x_2 - x_1) + 5\cos 10t,$$
$$m_2x_2'' = - 10(x_2 - x_1).$$

When we substitute $x_1 = A\cos 10t$, $x_2 = B\cos 10t$ and
cancel $\cos 10t$ throughout we get the equations

$$-40A - \qquad\qquad 10B = 5$$
$$-10A + (10 - 100m_2)B = 0.$$

If $m_2 = 0.1$ (slug)   then it follows that $A = 0$,   so the
mass m_1 remains at rest.

13.   First we need the general solution of the homogeneous system
$x'' = Ax$ with

$$A = \begin{bmatrix} -50 & 25/2 \\ 50 & -50 \end{bmatrix}.$$

The eigenvalues of A are $\lambda_1 = -25$ and $\lambda_2 = -75$, so the natural frequencies of the system are $\omega_1 = 5$ and $\omega_2 = 5\sqrt{3}$. The associated eigenvectors are $v_1 = [1 \quad 2]^T$ and $v_2 = [1 \quad -2]^T$, so the complementary solution $x_c(t)$ is given by

$$x_1(t) = c_1\cos 5t + c_2\sin 5t + c_3\cos 5\sqrt{3}t + c_4\sin 5\sqrt{3}t$$
$$x_2(t) = 2c_1\cos 5t + 2c_2\sin 5t - 2c_3\cos 5\sqrt{3}t - 2c_4\sin 5\sqrt{3}t.$$

When we substitute the trial solution $x_p(t) = [a \quad b]^T\cos 10t$ in the nonhomogeneous system, we find that $a = 4/3$, $b = -16/3$, so a particular solution $x_p(t)$ is described by

$$x_1(t) = (4/3)\cos 10t, \quad x_2(t) = -(16/3)\cos 10t.$$

Finally, when we impose the zero initial conditions on the solution $x(t) = x_c(t) + x_p(t)$ we find that $c_1 = 2/3$, $c_2 = 0$, $c_3 = -2$, $c_4 = 0$. Thus the solution we seek is described by

$$x_1(t) = (2/3)\cos 5t - 2\cos 5\sqrt{3}t + (4/3)\cos 10t$$
$$x_2(t) = (4/3)\cos 5t + 4\cos 5\sqrt{3}t + (16/3)\cos 10t.$$

We have a superposition of two oscillations with the natural frequencies $\omega_1 = 5$ and $\omega_2 = 5\sqrt{3}$ and a forced oscillation with frequency $\omega = 10$. In each of the two natural oscillations the amplitude of motion of m_2 is twice that of m_1, while in the forced oscillation the amplitude of motion of m_2 is four times that of m_1.

14. The characteristic equation of A is

$$(-c_1 - \lambda)(-c_2 - \lambda) - c_1c_2 = \lambda^2 + (c_1 + c_2)\lambda = 0,$$

whence the given eigenvalues and eigenvectors follow readily.

15. With $c_1 = c_2 = 2$, it follows from Problem 14 that the natural frequencies and associated eigenvectors are $\omega_1 = 0$, $v_1 = [1 \quad 1]^T$ and $\omega_2 = 2$, $v_2 = [1 \quad -1]^T$. Hence Theorem 1 gives the general solution

$$x_1(t) = a_1 + b_1 t + a_2 \cos 2t + b_2 \sin 2t$$
$$x_2(t) = a_1 + b_1 t - a_2 \cos 2t - b_2 \sin 2t.$$

The initial conditions $x_1'(0) = v_0$, $x_1(0) = x_2(0) = x_2'(0) = 0$ yield $a_1 = a_2 = 0$ and $b_1 = v_0/2$, $b_2 = v_0/4$, so

$$x_1(t) = (v_0/4)(2t + \sin 2t)$$
$$x_2(t) = (v_0/4)(2t - \sin 2t)$$

while $x_2 - x_1 = (v_0/4)(-2 \sin 2t) < 0$, that is, until $t = \pi/2$. Finally, $x_1'(\pi/2) = 0$ and $x_2'(\pi/2) = v_0$.

16. With $c_1 = 6$ and $c_2 = 3$, it follows from Problem 14 that the natural frequencies and associated eigenvectors are $\omega_1 = 0$, $v_1 = [1 \quad 1]^T$ and $\omega_2 = 3$, $v_2 = [2 \quad -1]^T$. Hence Theorem 1 gives the general solution

$$x_1(t) = a_1 + b_1 t + 2a_2 \cos 3t + 2b_2 \sin 3t$$
$$x_2(t) = a_1 + b_1 t - a_2 \cos 3t - b_2 \sin 3t.$$

The initial conditions $x_1'(0) = v_0$, $x_1(0) = x_2(0) = x_2'(0) = 0$ yield $a_1 = a_2 = 0$ and $b_1 = v_0/3$, $b_2 = v_0/9$, so

$$x_1(t) = (v_0/9)(3t + 2\sin 3t)$$
$$x_2(t) = (v_0/9)(3t - \sin 3t)$$

while $x_2 - x_1 = (v_0/9)(-3 \sin 3t) < 0$, that is, until $t = \pi/3$. Finally, $x_1'(\pi/3) = -v_0/3$ and $x_2'(\pi/3) = 2v_0/3$.

17. With $c_1 = 1$ and $c_2 = 3$, it follows from Problem 14 that the natural frequencies and associated eigenvectors are $\omega_1 = 0$, $v_1 = [1 \quad 1]^T$ and $\omega_2 = 2$, $v_2 = [1 \quad -3]^T$. Hence Theorem 1 gives the general solution

$$x_1(t) = a_1 + b_1 t + a_2 \cos 2t + b_2 \sin 2t$$
$$x_2(t) = a_1 + b_1 t - 3a_2 \cos 2t - 3b_2 \sin 2t.$$

The initial conditions $x_1'(0) = v_0$, $x_1(0) = x_2(0) = x_2'(0) = 0$ yield $a_1 = a_2 = 0$ and $b_1 = 3v_0/4$, $b_2 = v_0/8$, so

$$x_1(t) = (v_0/8)(6t + \sin 2t)$$
$$x_2(t) = (v_0/8)(6t - 3 \sin 2t)$$

while $x_2 - x_1 = (v_0/8)(-4 \sin 2t) < 0$, that is, until $t = \pi/2$. Finally, $x_1'(\pi/2) = v_0/2$ and $x_2'(\pi/2) = 3v_0/2$.

18. With $c_1 = c_3 = 4$ and $c_2 = 16$ the characteristic equation of the matrix

$$A = \begin{bmatrix} -4 & 4 & 0 \\ 16 & -32 & 16 \\ 0 & 4 & -4 \end{bmatrix}$$

is

$$\lambda^3 + 40\lambda^2 + 144\lambda = \lambda(\lambda + 4)(\lambda + 36) = 0.$$

The resulting eigenvalues, natural frequencies, and associated eigenvectors are

$$\lambda_1 = 0, \quad \omega_1 = 0, \quad v_1 = [1 \quad 1 \quad 1]^T$$
$$\lambda_2 = -4, \quad \omega_2 = 2, \quad v_2 = [1 \quad 0 \quad -1]^T$$
$$\lambda_3 = -36, \quad \omega_3 = 6, \quad v_3 = [1 \quad -8 \quad 1]^T$$

Theorem 1 then gives the general solution

$$x_1(t) = a_1 + b_1 t + a_2 \cos 2t + b_2 \sin 2t$$
$$+ a_3 \cos 6t + b_3 \sin 6t$$
$$x_2(t) = a_1 + b_1 t - 8a_3 \cos 6t - 8b_3 \sin 6t$$
$$x_3(t) = a_1 + b_1 t - a_2 \cos 2t - b_2 \sin 2t$$
$$+ a_3 \cos 6t + b_3 \sin 6t$$

The initial conditons yield $a_1 = a_2 = a_3 = 0$ and $b_1 = 4v_0/9$, $b_2 = v_0/4$, $b_3 = v_0/108$, so

$$x_1(t) = (v_0/108)(48t + 27 \sin 2t + \sin 6t)$$
$$x_2(t) = (v_0/108)(48t - 8 \sin 6t)$$
$$x_3(t) = (v_0/108)(48t - 27 \sin 2t + \sin 6t)$$

while

$$x_2 - x_1 = -18(\sin 2t)(3 - 2 \sin^3 2t) < 0,$$
$$x_3 - x_2 = -9(4 \sin^3 2t) < 0,$$

that is, until $t = \pi/2$. Finally

$$x_1'(\pi/2) = -v_0/9, \quad x_2(\pi/2) = 8v_0/9, \quad x_3'(\pi/2) = 8v_0/9.$$

19. The matrix

$$A = \begin{bmatrix} -20 & 10 \\ 10 & -10 \end{bmatrix}$$

has characteristic equation $\lambda^2 - 30\lambda + 100 = 0$ with roots $\lambda_1 = -26.1803$ and $\lambda_2 = -3.8197$. Hence the natural frequencies are $\omega_1 = 5.1167$ rad/sec, $\omega_2 = 1.9544$ rad/sec and the natural periods are $P_1 = 2\pi/\omega_1 = 1.2280$ sec, $P_2 = 2\pi/\omega_2 = 3.2149$ sec.

20. The matrix

$$A = \begin{bmatrix} -20 & 10 & 0 \\ 10 & -20 & 16 \\ 0 & 10 & -10 \end{bmatrix}$$

has characteristic equation

$$\lambda^3 + 50\lambda^2 + 600\lambda + 1000 = 0$$

whose roots (found by Newton's method, or whatever) are λ_1 = -32.4698, λ_2 = -15.5496, and λ_3 = -1.9806. Hence the natural frequencies are ω_1 = 5.6982, ω_2 = 3.9433, and ω_3 = 1.4073. The natural periods are P_1 = 1.1027 sec, P_2 = 1.5934 sec, and P_3 = 4.4646 sec.

21. The matrix

$$A = \begin{bmatrix} -20 & 10 & 0 \\ 20 & -40 & 20 \\ 0 & 20 & -20 \end{bmatrix}$$

has characteristic equation

$$\lambda^3 + 80\lambda^2 + 1400\lambda + 4000 = (\lambda + 20)(\lambda^2 + 60\lambda + 200) = 0$$

so the eigenvalues are λ_1 = -20, λ_2, λ_3 = -30 ± 10$\sqrt{7}$. The associated eigenvectors are v_1 = [1 0 -1]^T and v_2, v_3 = [1 -1 ± $\sqrt{7}$ 2]^T.

22. A period of 3 seconds corresponds to circular frequency ω = $2\pi/3$, and the amplitude is E = 1/4 ft. Inserting terms of the form F = $mE\omega^2$sin ωt in each of the equations of Problem 15, and dividing each equation by the mass, we get the system

$$x'' + 20x - 10y \qquad = (1/4)\omega^2 \sin \omega t,$$
$$y'' - 20x + 40y - 20z = (1/4)\omega^2 \sin \omega t,$$
$$z'' \qquad - 20y + 20z = (1/4)\omega^2 \sin \omega t.$$

When we substitute x = A sin ωt, y = B sin ωt, and z = C sin ωt, we find that the amplitudes are

$$A \approx -0.8185 \text{ ft} \approx -9.82 \text{ in},$$
$$B \approx -1.3877 \text{ ft} \approx -16.65 \text{ in},$$
$$C \approx -1.7073 \text{ ft} \approx -20.49 \text{ in}.$$

23. The matrix

$$A = \begin{bmatrix} -160/3 & 320/3 \\ 8 & -116 \end{bmatrix}$$

has eigenvalues $\lambda_1 \approx -41.8285$ and $\lambda_2 \approx -127.5049$, so the natural frequencies are

$$\omega_1 \approx 6.4675 \text{ rad/sec} \approx 1.0293 \text{ Hz}$$
$$\omega_2 \approx 11.2918 \text{ rad/sec} \approx 1.7971 \text{ Hz}.$$

(b) Resonance occurs at the two critical speeds

$$v_1 = 20\omega_1/\pi \approx 41 \text{ ft/sec} \approx 28 \text{ mi/h},$$
$$v_2 = 20\omega_2/\pi \approx 72 \text{ ft/sec} \approx 49 \text{ mi/h}$$

24. With $k_1 = k_2 = k$ and $L_1 = L_2 = L/2$ the equations in (42) reduce to

$$mx" = -2kx \quad \text{and} \quad I\theta" = -kL^2\theta/2.$$

The first equation yields $\omega_1 = \sqrt{(2k/m)}$ and the second one yields $\omega_2 = \sqrt{(kL^2/2I)}$.

In Problems 25-27 we substitute the given physical parameters into the equations in (42):

$$mx" = -(k_1 + k_2)x + (k_1L_1 - k_2L_2)\theta$$

$$I\theta" = (k_1L_1 - k_2L_2)x - (k_1L_1^2 + k_2L_2^2)\theta$$

As in Problem 23, a critical frequency of ω rad/sec yields a critical velocity of $v = 20\omega/\pi$ ft/sec.

25. $100x" + 4000x = 0$, $800\theta" + 100000\theta = 0$

Up-and-down: $\omega_1 = \sqrt{(40)}$, $v_1 \approx 40.26$ ft/sec ≈ 27 mph

Angular: $\omega_2 = \sqrt{(125)}$, $v_2 \approx 71.18$ ft/sec ≈ 49 mph

26. $100x'' + 4000x - 4000\theta = 0$

$1000\theta'' + 4000x + 104000\theta = 0$

$\omega_1 \approx 6.1311,\quad v_1 \approx 39.03$ ft/sec ≈ 27 mph

$\omega_2 \approx 10.3155,\quad v_2 \approx 65.67$ ft/sec ≈ 45 mph

27. $100x'' + 3000x + 5000\theta = 0$

$800\theta'' + 5000x + 75000\theta = 0$

$\omega_1 \approx 5.0424,\quad v_1 \approx 32.10$ ft/sec ≈ 22 mph

$\omega_2 \approx 9.9158,\quad v_2 \approx 63.13$ ft/sec ≈ 43 mph

SECTION 5.6

MULTIPLE EIGENVALUE SOLUTIONS

In each of Problems 1-6 we give the characteristic equation with repeated eigenvalue λ, the single eigenvector v and a generalized eigenvector w such that $(A - \lambda I)w = v$, and the scalar component functions $x_1(t)$, $x_2(t)$ of the general solution

$$x(t) = c_1 v e^{\lambda t} + c_2 (vt + w) e^{\lambda t}$$

of the given system $x' = Ax$.

1. Characteristic equation $\lambda^2 + 6\lambda + 9 = 0$

Repeated eigenvalue $\lambda = -3$

Eigenvector $v = [1 \quad -1]^T$.

Generalized eigenvector $w = [1 \quad 0]^T$

$$x_1(t) = (c_1 + c_2 + c_2 t) e^{-3t}$$
$$x_2(t) = (-c_1 \qquad - c_2 t) e^{-3t}.$$

2. Characteristic equation $\lambda^2 - 4\lambda + 4 = 0$

Repeated eigenvalue $\lambda = 2$

Eigenvector $v = [1 \quad 1]^T$.

Generalized eigenvector $w = [1 \quad 0]^T$

$$x_1(t) = (c_1 + c_2 + c_2 t) e^{2t}$$
$$x_2(t) = (c_1 + \qquad c_2 t) e^{2t}.$$

3. Characteristic equation $\lambda^2 - 6\lambda + 9 = 0$

Repeated eigenvalue $\lambda = 3$

Single eigenvector $v = [-2 \quad 2]^T$.

Generalized eigenvector $w = [1 \quad 0]^T$

$$x_1(t) = (-2c_1 + c_2 - 2c_2 t) e^{3t}$$
$$x_2(t) = (2c_1 + \qquad 2c_2 t) e^{3t}.$$

4. Characteristic equation $\lambda^2 - 8\lambda + 16 = 0$

Repeated eigenvalue $\lambda = 4$

Single eigenvector $v = [-1 \quad 1]^T$

Generalized eigenvector $w = [1 \quad 0]^T$

$$x_1(t) = (-c_1 + c_2 - c_2t)e^{4t}$$
$$x_2(t) = (\ c_1 + \qquad c_2t)e^{4t}.$$

5. Characteristic equation $\lambda^2 - 10\lambda + 25 = 0$

Repeated eigenvalue $\lambda = 5$

Single eigenvector $v = [2 \quad -4]^T$

Generalized eigenvector $w = [1 \quad 0]^T$

$$x_1(t) = (\ 2c_1 + c_2 + 2c_2t)e^{5t}$$
$$x_2(t) = (-4c_1 \qquad - 4c_2t)e^{5t}.$$

6. Characteristic equation $\lambda^2 - 10\lambda + 25 = 0$

Repeated eigenvalue $\lambda = 5$

Single eigenvector $v = [-4 \quad -4]^T$

Generalized eigenvector $w = [1 \quad 0]^T$

$$x_1(t) = (-4c_1 + c_2 - 4c_2t)e^{5t}$$
$$x_2(t) = (\ 4c_1 \qquad + 4c_2t)e^{5t}.$$

In each of Problems 7-10 the characteristic polynomial is easily calculated by expansion along the row or column of   A   that contains two zeros. The matrix   A   has a double eigenvalue, but nevertheless has 3 linearly independent eigenvectors. We give also the scalar components $x_1(t)$, $x_2(t)$, $x_3(t)$ of the general solution of the system.

7. Characteristic equation $(\lambda - 2)^2(\lambda - 9) = 0$

Eigenvalues $\lambda = 2, \quad 2, \quad 9$

Eigenvectors $[1 \quad 1 \quad 0]^T, \quad [1 \quad 0 \quad 1]^T, \quad [0 \quad 1 \quad 0]^T$

$$x_1(t) = c_1 e^{2t} + c_2 e^{2t}$$
$$x_2(t) = c_1 e^{2t} \qquad\qquad + c_3 e^{9t}$$
$$x_3(t) = \qquad\qquad c_2 e^{2t}$$

8. Characteristic equation $(\lambda - 7)(\lambda - 13)^2 = 0$

Eigenvalues $\lambda = 7, \quad 13, \quad 13$

Eignevectors $[2 \quad -3 \quad 1]^T, \quad [1 \quad -1 \quad 0]^T, \quad [0 \quad 0 \quad 1]^T$

$$x_1(t) = 2c_1 e^{7t} + c_2 e^{13t}$$
$$x_2(t) = -3c_1 e^{7t} - c_2 e^{13t}$$
$$x_3(t) = c_1 e^{7t} \qquad\qquad + c_3 e^{13t}$$

9. Characteristic equation $(\lambda - 5)^2(\lambda - 9) = 0$

Eigenvalues $\lambda = 5, \quad 5, \quad 9$

Eigenvectors $[1 \quad 2 \quad 0]^T, \quad [7 \quad 0 \quad 2]^T, \quad [3 \quad 0 \quad 1]^T$

$$x_1(t) = c_1 e^{5t} + 7c_2 e^{5t} + 3c_3 e^{9t}$$
$$x_2(t) = 2c_1 e^{5t}$$
$$x_3(t) = \qquad\qquad 2c_2 e^{5t} + c_3 e^{9t}$$

10. Characteristic equation $(\lambda - 3)^2(\lambda - 7) = 0$

Eigenvalues $\lambda = 3, \quad 3, \quad 7$

Eigenvectors $[5 \quad 2 \quad 0]^T$, $[3 \quad 0 \quad -1]^T$, $[2 \quad 1 \quad 0]^T$

$$x_1(t) = 5c_1e^{3t} + 3c_2e^{3t} + 2c_3e^{7t}$$
$$x_2(t) = 2c_1e^{3t} \qquad\qquad + c_3e^{7t}$$
$$x_3(t) = \qquad\qquad - c_2e^{3t}$$

In each of Problems 11-14, $\lambda = -1$ is a triple eigenvalue of defect 2, and we give a length 2 chain $\{v_1, v_2, v_3\}$ of generalized eigenvectors. The general solution then is given by

$$x(t) = e^{-t}[c_1v_1 + c_2(v_1t + v_2) + c_3(v_1t^2/2 + v_2t + v_3)].$$

We give the scalar components $x_1(t)$, $x_2(t)$, $x_3(t)$ of $x(t)$.

11. $v_1 = [0 \quad 1 \quad 0]^T$, $\qquad v_2 = [-2 \quad -1 \quad 1]^T$, $\qquad v_3 = [1 \quad 0 \quad 0]^T$

$$x_1(t) = e^{-t}(-2c_2 + c_3 - 2c_3t)$$
$$x_2(t) = e^{-t}(c_1 - c_2 + c_2t - c_3t + c_3t^2/2)$$
$$x_3(t) = e^{-t}(c_2 + c_3t)$$

12. $v_1 = [1 \quad 1 \quad 0]^T$, $\qquad v_2 = [0 \quad 0 \quad 1]^T$, $\qquad v_3 = [1 \quad 1 \quad 0]^T$

$$x_1(t) = e^{-t}(c_1 + c_3 + c_2t + c_3t^2/2)$$
$$x_2(t) = e^{-t}(c_1 + c_2t + c_3t^2/2)$$
$$x_3(t) = e^{-t}(c_2 + c_3t)$$

13. $v_1 = [1 \quad 0 \quad 0]^T$, $\qquad v_2 = [0 \quad 2 \quad 1]^T$, $\qquad v_3 = [0 \quad 1 \quad 0]^T$

$$x_1(t) = e^{-t}(c_1 + c_2t + c_3t^2/2)$$
$$x_2(t) = e^{-t}(2c_2 + c_3 + 2c_2t)$$
$$x_3(t) = e^{-t}(c_2 + c_3t)$$

14. $v_1 = [5 \quad -25 \quad -5]^T, \quad v_2 = [1 \quad -5 \quad 4]^T, \quad v_3 = [\ 1 \quad 0 \quad 0]^T$

$$x_1(t) = e^{-t}(5c_1 + c_2 + c_3 + 5c_2t + c_3t + 5c_3t^2/2)$$
$$x_2(t) = e^{-t}(-25c_1 - 5c_2 - 25c_2t - 5c_3t - 25c_3t^2/2)$$
$$x_3(t) = e^{-t}(-5c_1 + 4c_2 - 5c_2t + 4c_3t - 5c_3t^2/2)$$

In each of Problems 15-18, $\lambda = 1$ is a triple eigenvalue of defect 1. We give two eigenvectors u_1 and u_2, and a length 2 chain $\{v_1, v_2\}$ of generalized eigenvectors. The general solution is then given by

$$x(t) = e^t[c_1u_1 + c_2u_2 + c_3(v_1t + v_2)]$$

We give the scalar components $x_1(t)$, $x_2(t)$, $x_3(t)$ of $x(t)$.

15. $u_1 = [3 \quad -1 \quad 0]^T \qquad u_2 = [0 \quad 0 \quad 1]^T$

$v_1 = [-3 \quad 1 \quad 1]^T \qquad v_2 = [1 \quad 0 \quad 0]^T$

$$x_1(t) = e^t(3c_1 - 3c_2 + c_3 - 3c_3t)$$
$$x_2(t) = e^t(-c_1 + c_2 + c_3t)$$
$$x_3(t) = e^t(c_2 + c_3t)$$

16. $u_1 = [3 \quad -2 \quad 0]^T \qquad u_2 = [3 \quad 0 \quad -2]^T$

$v_1 = [0 \quad -2 \quad 2]^T \qquad v_2 = [1 \quad 0 \quad 0]^T$

$$x_1(t) = e^t(3c_1 + c_3)$$

$$x_2(t) = e^t(-2c_1 - 2c_2 - 2c_3t)$$

$$x_3(t) = e^t(2c_2 + 2c_3t)$$

17. $u_1 = [2 \quad 0 \quad -9]^T$ $\qquad u_2 = [1 \quad -3 \quad 0]^T$

$v_1 = [0 \quad 6 \quad -9]^T$ $\qquad v_2 = [0 \quad 1 \quad 0]^T$

$$x_1(t) = e^t(2c_1)$$

$$x_2(t) = e^t(6c_2 + c_3 + 6c_3t)$$

$$x_3(t) = e^t(-9c_1 - 9c_2 - 9c_3t)$$

18. $u_1 = [-1 \quad 0 \quad 1]^T$ $\qquad u_2 = [-2 \quad 1 \quad 0]^T$

$v_1 = [0 \quad 1 \quad -2]^T$ $\qquad v_2 = [1 \quad 0 \quad 0]^T$

$$x_1(t) = e^t(-c_1 + c_3)$$

$$x_2(t) = e^t(c_2 + c_2t)$$

$$x_3(t) = e^t(c_1 - 2c_2 - 2c_2t)$$

19. Characteristic equation $\lambda^4 - 2\lambda^2 + 1 = 0$

Double eigenvalue $\lambda = -1$ with eigenvectors
$v_1 = [1 \quad 0 \quad 0 \quad 1]^T$ and $v_2 = [0 \quad 0 \quad 1 \quad 0]^T$..

Double eigenvalue $\lambda + 1$ with eigenvectors
$v_3 = [0 \quad 1 \quad 0 \quad -2]^T$ and $v_4 = [1 \quad 0 \quad 3 \quad 0]^T$.

General solution
$$x(t) = e^{-t}(c_1v_1 + c_2v_2) + e^t(c_3v_3 + c_4v_4)$$

Scalar components

$$x_1(t) = c_1 e^{-t} + c_4 e^t$$
$$x_2(t) = c_3 e^t$$
$$x_3(t) = c_2 e^{-t} + 3 c_4 e^t$$
$$x_4(t) = c_1 e^{-t} - 2 c_3 e^t$$

20. Characteristic equation $(\lambda - 2)^4 = 0$

Eigenvalue $\lambda = 2$ with multiplicity 4 and defect 3.

Length 4 chain $\{v_1, v_2, v_3, v_4\}$ with

$$v_1 = [1 \quad 0 \quad 0 \quad 0]^T \qquad v_2 = [0 \quad 1 \quad 0 \quad 0]^T$$
$$v_3 = [1 \quad 0 \quad 1 \quad 0]^T \qquad v_4 = [0 \quad 0 \quad 0 \quad 1]^T$$

General solution

$$x(t) = e^t [c_1 v_1 + c_2(v_1 t + v_2) + c_3(v_1 t^2/2 + v_2 t + v_3)$$
$$+ c_4(v_1 t^3/6 + v_2 t^2/2 + v_3 t \quad v_4)]$$

Scalar components

$$x_1(t) = e^{2t}(c_1 + c_3 + c_2 t + c_4 t + c_3 t^2/2 + c_4 t^3/6)$$
$$x_2(t) = e^{2t}(c_2 + c_3 t + c_4 t^2/2)$$
$$x_3(t) = e^{2t}(c_3 + c_4 t)$$
$$x_4(t) = e^{2t}(c_4)$$

21. Characteristic equation $(\lambda - 1)^4 = 0$

Eigenvalue $\lambda = 1$ with multiplicity 4 and defect 2.

Chains $\{v_1, v_2, v_3\}$ and $\{v_4\}$ with

$$v_1 = [0 \quad 0 \quad 0 \quad 1]^T \qquad v_2 = [-2 \quad 1 \quad 1 \quad 0]^T$$

$$v_3 = [1 \quad 0 \quad 0 \quad 0]^T \qquad v_4 = [0 \quad 0 \quad 1 \quad 0]^T$$

General solution

$$x(t) = e^{-t}[c_1 v_1 + c_2(v_1 t + v_2) + c_3(v_1 t^2/2 + v_2 t + v_3) + c_4 v_4]$$

Scalar components

$$x_1(t) = e^t(-2c_2 + c_3 - 2c_2 t)$$

$$x_2(t) = e^t(c_2 + c_3 t)$$

$$x_3(t) = e^t(c_2 + c_4 + c_3 t)$$

$$x_4(t) = e^t(c_1 + c_2 + c_3 t^2/2)$$

22. Same eigenvalue and chain structure as in Problem 21, but with generalized eigenvectors

$$v_1 = [1 \quad 0 \quad 0 \quad -2]^T \qquad v_2 = [3 \quad -2 \quad 1 \quad -6]^T$$

$$v_3 = [0 \quad 1 \quad 0 \quad 0]^T \qquad v_4 = [1 \quad 0 \quad 0 \quad 0]^T$$

Scalar components

$$x_1(t) = e^t(c_1 + 3c_2 + c_4 + c_2 t + 3c_3 t + c_3 t^2/2)$$

$$x_2(t) = e^t(-2c_2 - 2c_3 t)$$

$$x_3(t) = e^t(c_2 + c_3 t)$$

$$x_4(t) = e^t(-2c_1 - 6c_2 - 2c_2 t - 6c_3 t - c_3 t^2)$$

In Problems 23-32 we give the generalized eigenvector chain(s) associated with each eigenvalue.

23. $\lambda = -1$: $\{v_1\}$ with $v_1 = [1 \quad -1 \quad 2]^T$

$\lambda = 3$: $\{v_2\}$ with $v_2 = [4 \quad 0 \quad 9]^T$ and

$\{v_3\}$ with $v_3 = [0 \quad 2 \quad 1]^T$

Scalar components

$$x_1(t) = c_1e^{-t} + 4c_2e^{3t}$$
$$x_2(t) = -c_1e^{-t} \qquad\qquad + 2c_3e^{3t}$$
$$x_3(t) = 2c_1e^{-t} + 9c_2e^{3t} + c_3e^{3t}$$

24. $\lambda = -2$: $\{v_1\}$ with $v_1 = [5 \quad 3 \quad -3]^T$

$\lambda = 3$: $\{v_2\}$ with $v_2 = [4 \quad 0 \quad -1]^T$ and

$\{v_3\}$ with $v_3 = [2 \quad -1 \quad 0]^T$

Scalar components

$$x_1(t) = 5e^{-2t} + 4c_2e^{3t} + 2c_3e^{3t}$$
$$x_2(t) = 3e^{-2t} \qquad\qquad - c_3e^{3t}$$
$$x_3(t) = -3e^{-2t} - c_2e^{3t}$$

25. $\{v_1, v_2, v_3\}$ with

$v_1 = [-1 \quad 0 \quad -1]^T$, $v_2 = [-4 \quad -1 \quad 0]^T$, $v_3 = [1 \quad 0 \quad 0]^T$

General solution

$$x(t) = c_1v_1e^{2t} + c_2(v_1t + v_2)e^{2t} + c_3(v_1t^2/2 + v_2t + v_3)e^{2t}$$

26. $\{v_1, v_2, v_3\}$ with

$v_1 = [0 \quad 2 \quad 2]^T$, $v_2 = [2 \quad 1 \quad -3]^T$, $v_3 = [1 \quad 0 \quad 0]^T$

General solution

$$x(t) = e^{3t}[c_1 v_1 + c_2(v_1 t + v_2) + c_3(v_1 t^2/2 + v_2 t + v_3)]$$

27. $\{v_1, v_2\}$ and $\{v_3\}$ with

$v_1 = [-5 \quad 3 \quad 8]^T$, $v_2 = [1 \quad 0 \quad 0]^T$ and $v_3 = [1 \quad 1 \quad 0]^T$.

General solution

$$x(t) = e^{2t}[c_1 v_1 + c_2(v_1 t + v_2) + c_3 v_3]$$

28. $\{v_1, v_2, v_3\}$ with

$v_1 = [119 \quad -289 \quad 0]^T$, $v_2 = [-17 \quad 34 \quad 17]^T$, $v_3 = [1 \quad 0 \quad 0]^T$

General solution

$$x(t) = e^{2t}[c_1 v_1 + c_2(v_1 t + v_2) + c_3(v_1 t^2/2 + v_2 t + v_3)]$$

29. $\lambda = -1$: $\{v_1, v_2\}$ with $v_1 = [1 \quad -3 \quad -1 \quad -2]^T$

and $v_2 = [0 \quad 1 \quad 0 \quad 0]^T$,

$\lambda = 2$: $\{u_1, u_2\}$ with $u_1 = [0 \quad -1 \quad 1 \quad 0]^T$

and $u_2 = [0 \quad 0 \quad 2 \quad 1]^T$

General solution

$$x(t) = e^{-t}[c_1 v_1 + c_2(v_1 t + v_2)] + e^{2t}[c_3 u_1 + c_4(u_1 t + u_2)]$$

30. $\lambda = -1$: $\{v_1, v_2\}$ with $v_1 = [0 \quad 1 \quad -1 \quad -3]^T$

and $v_2 = [0 \quad 0 \quad 1 \quad 2]^T$,

$\lambda = 2$: $\{u_1, u_2\}$ with $u_1 = [-1 \quad 0 \quad 0 \quad 0]^T$

and $u_2 = [0 \quad 0 \quad 3 \quad 5]^T$

General solution

$$x(t) = e^{-t}[c_1 v_1 + c_2(v_1 t + v_2)] + e^{2t}[c_3 u_1 + c_4(u_1 t + u_2)]$$

31. $\{v_1, v_2, v_3\}$ and $\{v_4\}$ with

$v_1 = [42 \quad 7 \quad -21 \quad -42]^T$, $\quad v_2 = [34 \quad 22 \quad -10 \quad -27]^T$,

$v_3 = [1 \quad 0 \quad 0 \quad 0]^T$ and $\quad v_4 = [0 \quad 1 \quad 3 \quad 0]$

General solution

$$x(t) = e^t[c_1 v_1 + c_2(v_1 t + v_2) + c_3(v_1 t^2/2 + v_2 t + v_3) + c_4 v_4]$$

32. $\lambda = 2$: eigenvectors $v_1 = [8 \quad 0 \quad -3 \quad 1 \quad 0]^T$ and

$$v_2 = [1 \quad 0 \quad 0 \quad 0 \quad 3]^T$$

$\lambda = 3$: eigenvectors $v_3 = [3 \quad -2 \quad -1 \quad 0 \quad 0]^T$,

$$v_4 = [2 \quad -2 \quad 0 \quad -3 \quad 0]^T, \quad \text{and}$$

$$v_5 = [1 \quad -1 \quad 0 \quad 0 \quad 3]^T$$

General solution

$$x(t) = e^{2t}(c_1 v_1 + c_2 v_2) + e^{3t}(c_3 v_3 + c_4 v_4 + c_5 v_5)$$

33. The chain $\{v_1, v_2\}$ was found using the matrices

$$A - \lambda I = \begin{bmatrix} 4i & -4 & 1 & 0 \\ 4 & 4i & 0 & 1 \\ 0 & 0 & 4i & -4 \\ 0 & 0 & 4 & 4i \end{bmatrix} \longrightarrow \begin{bmatrix} 1 & i & 0 & 0 \\ 0 & 0 & 1 & 0 \\ 0 & 0 & 0 & 1 \\ 0 & 0 & 0 & 0 \end{bmatrix}$$

and

$$(A - \lambda I)^2 = \begin{bmatrix} -32 & -32i & 8i & -8 \\ 32i & -32 & 8 & 8i \\ 0 & 0 & -32 & -32i \\ 0 & 0 & 32i & -32 \end{bmatrix} \longrightarrow \begin{bmatrix} 1 & i & 0 & 0 \\ 0 & 0 & 1 & i \\ 0 & 0 & 0 & 0 \\ 0 & 0 & 0 & 0 \end{bmatrix}$$

where $\longrightarrow$ signifies reduction to row-echelon form. The resulting real-valued solution vectors are

$$x_1(t) = e^{3t}[\quad \cos 4t \qquad \sin 4t \qquad 0 \qquad 0\ ]^T$$

$$x_2(t) = e^{3t}[\ -\sin 4t \qquad \cos 4t \qquad 0 \qquad 0]^T$$

$$x_3(t) = e^{3t}[\quad t\cos 4t \quad t\sin 4t \quad \cos 4t \quad \sin 4t]^T$$

$$x_4(t) = e^{3t}[\ -t\sin 4t \quad t\cos 4t \quad -\sin 4t \quad \cos 4t]^T$$

34. The chain $\{v_1, v_2\}$ was found using the matrices

$$A - \lambda I = \begin{bmatrix} 3i & 0 & -8 & -3 \\ -18 & -3-3i & 0 & 0 \\ -9 & -3 & -27-3i & -9 \\ 33 & 10 & 90 & 30-3i \end{bmatrix}$$

$$\longrightarrow \begin{bmatrix} 1 & 0 & 0 & 0 \\ 0 & 1 & 0 & 3+3i \\ 0 & 0 & 1 & 0 \\ 0 & 0 & 0 & 0 \end{bmatrix}$$

and

$$(A - \lambda I)^2 = \begin{bmatrix} -36 & -6 & -54+48i & -18+18i \\ 54+108i & 18i & 144 & 54 \\ 54i & 18i & -18+162i & 54i \\ -198i & -60i & 6-540i & -18-180i \end{bmatrix}$$

$$\longrightarrow \begin{bmatrix} 1 & 0 & -3i & -i \\ 0 & 1 & 9+10i & 3+3i \\ 0 & 0 & 0 & 0 \\ 0 & 0 & 0 & 0 \end{bmatrix}$$

where —> signifies reduction to row-echelon form. The resulting real-valued solution vectors are

$$x_1(t) = e^{2t}[\; \sin 3t \quad\quad 3\cos 3t - 3\sin 3t \quad\quad 0 \quad\quad \sin 3t]^T$$

$$x_2(t) = e^{2t}[-\cos 3t \quad\quad 3\sin 3t + 3\cos 3t \quad\quad 0 \quad\quad -\cos 3t]^T$$

$$x_3(t) = e^{2t}[\; 3\cos 3t + t\sin 3t \quad (3t - 10)\cos 3t - (3t + 9)\sin 3t$$
$$\sin 3t \quad\quad\quad\quad t\sin 3t \;]^T$$

$$x_4(t) = e^{2t}[-t\cos 3t + 3\sin 3t \quad (3t + 9)\cos 3t + (3t - 10)\sin 3t$$
$$-\cos 3t \quad\quad\quad\quad -t\cos 3t \;]^T$$

35. The coefficient matrix

$$A = \begin{bmatrix} 0 & 0 & 1 & 0 \\ 0 & 0 & 0 & 1 \\ -1 & 1 & -2 & 1 \\ 1 & -1 & 1 & -2 \end{bmatrix}$$

has eigenvalues

$$\lambda = 0 \quad \text{with eigenvector} \quad v_1 = [1 \quad 1 \quad 0 \quad 0]^T$$

$$\lambda = -1 \quad \text{with eigenvectors} \quad v_2 = [1 \quad 0 \quad -1 \quad 0]^T$$
$$\text{and} \quad v_3 = [0 \quad 1 \quad 0 \quad -1]^T,$$

$$\lambda = -2 \quad \text{with eigenvector} \quad v_4 = [1 \quad -1 \quad -2 \quad 2]^T.$$

When we impose the given initial condiitions on the general solution

$$x(t) = c_1 v_1 + c_2 v_2 e^{-t} + c_3 v_3 e^{-t} + c_4 v_4 e^{-2t}$$

We find that $c_1 = v_0$, $c_2 = c_3 = -v_0$, $c_4 = 0$. Hence the position functions of the two masses are given by

$$x_1(t) = x_2(t) = v_0(1 - e^{-t})$$

Each mass travels a distance v_0 before stopping.

36. The coefficient matrix is the same as in Problem 35 except that $a_{44} = -1$. Now the matrix A has the eigenvalue $\lambda = 0$ with eigenvector $v_0 = [1 \quad 1 \quad 0 \quad 0]^T$, and the triple eigenvalue $\lambda = -1$ with associated length 2 chain $\{v_1, v_2, v_3\}$ consisting of the generalized eigenvectors

$$v_1 = [0 \quad 1 \quad 0 \quad -1]^T$$
$$v_2 = [1 \quad 0 \quad -1 \quad 1]^T$$
$$v_3 = [1 \quad 0 \quad 0 \quad 0]^T.$$

When we impose the given initial conditions on the general solution

$$x(t) = c_0 v_0 + e^{-t}[c_1 v_1 + c_2(v_1 t + v_2) + c_3(v_1 t^2/2 + v_2 t + v_3)]$$

We find that $c_0 = 2v_0$, $c_1 = -2v_0$, $c_2 = c_3 = -v_0$. Hence the position functions of the two masses are given by

$$x_1(t) = v_0(2 - 2e^{-t} - te^{-t}),$$
$$x_2(t) = v_0(2 - 2e^{-t} - te^{-t} - t^2 e^{-t}/2).$$

Each travels a distance $2v_0$ before stopping.

NONHOMOGENEOUS LINEAR SYSTEMS

1. $x = 7/3$, $y = -8/3$

2.   The trial solution $x = a_1 + b_1t$, $y = a_2 + b_2t$ yields the particular solution

$$x = (1 + 12t)/8, \qquad y = -(5 + 4t)/4$$

3. $x = (864e^{-t} + 4e^{6t} - 504t^2 + 840t - 868)/756$
$y = (-864e^{-t} + 3e^{6t} + 378t^2 - 882t + 861)/756$

4.   By the eigenvalue method we find first the complementary solution

$$x_c = c_1e^{5t} + c_2e^{-2t}, \qquad y_c = c_1e^{5t} - 6c_2e^{-2t}.$$

Then we try

$$x = a_1e^t, \quad y = a_2e^t$$

to find the particular solution

$$x_p = -e^{-t}/12, \qquad y_p = -3e^t/4.$$

Thus the general solution is

$$x = c_1e^{5t} + c_2e^{-2t} - e^t/12,$$
$$y = c_1e^{5t} - 6c_2e^{-2t} - 3e^t/4.$$

Finally we apply the initial conditions $x(0) = y(0) = 1$ to determine c_1 and c_2. The result is

$$x = (99e^{5t} - 8e^{-2t} - 7e^t)/84,$$
$$y = (99e^{5t} + 48e^{-2t} - 63e^t)/84.$$

5. $x = (-12 - e^{-t} - 7te^{-t})/3,$ $y = (-6 - 7te^{-t})/3$

6. Substitution of the trial solution

$$x = a_1e^t + b_1te^t,   y = a_2e^t + b_2te^t$$

yields the particular solution

$$x = -e^t(91 + 16t)/256,   y = e^t(25 + 16t)/32.$$

7. $x = (369e^t + 166e^{-9t} - 125 \cos t - 105 \sin t)/410$
    $y = (369e^t - 249e^{-9t} - 120 \cos t - 150 \sin t)/410$

8. Substitution of the trial solution

$$x = a_1\cos t + b_1\sin t,   y = a_2\cos t + b_2\sin t$$

yields the particular solution

$$x = (17 \cos t + 2 \sin t)/3,$$
$$y = (3 \cos t + 5 \sin t)/3.$$

9. $x = (\sin 2t + 2t \cos 2t + t \sin 2t)/4$
    $y = (t \sin 2t)/4$

10. Substitution of the trial solution

$$x = a_1e^t\cos t + b_1e^t\sin t,$$
$$y = a_2e^t\cos t + b_2e^t\sin t$$

yields the particular solution

$$x = e^t(4 \cos t - 6 \sin t)/13,$$

$$y = e^t(3 \cos t + 2 \sin t)/13.$$

11. $x = (1 - 4t + e^{4t})/2, \qquad y = (-5 + 4t + e^{4t})/4$

12. Because $\lambda = 0$ is an eigenvalue of the associated homogeneous system, we substitute the trial solution

$$x = a_1 + b_1 t + c_1 t^2, \qquad y = a_2 + b_2 t + c_2 t^2.$$

The resulting equations

$$
\begin{aligned}
b_1 &= a_1 + a_2 \\
b_2 &= a_1 + a_2 \\
2c_1 &= b_1 + b_1 + 2 \\
2c_2 &= b_1 + b_2 - 2 \\
c_2 &= -c_1
\end{aligned}
$$

are satisfied by

$$a_1 = b_1 = c_1 = c_2 = 0, \quad c_1 = 1, \quad c_2 = -1.$$

This gives the particular solution $x = t^2, \quad y = -t^2.$

13. $x = (1 + 5t)e^t/2, \qquad y = -5te^t/2$

14. Because the associated homogeneous system has eigenvalues $\lambda = 0$ and $\lambda = 4$ we substitute the trial solution

$$
\begin{aligned}
x &= a_1 + b_1 t + c_1 e^{4t} + d_1 te^{4t}, \\
y &= a_2 + a_2 t + c_2 e^{4t} + d_2 te^{4t}.
\end{aligned}
$$

Upon solving the resulting eight equations in eight unknowns we obtain the particular solution

$$
\begin{aligned}
x &= (-1 + 2t + te^{4t})/4, \\
y &= (-4t + e^{4t} + 2te^{4t})/4.
\end{aligned}
$$

In Problems 15 - 22 we use the formula

$$x(t) = \Phi(t)\Phi(a)^{-1}b,$$

where $\Phi(t)$ is a fundamental matrix for the homogeneous system $x' = Px$, to find the solution vector $x(t)$ that satisfies the initial condition $x(a) = b$. Formulas (25) and (26) in the text provide inverses of 2-by-2 and 3-by-3 matrices.

15. $\varphi(t) = \begin{bmatrix} e^t & e^{3t} \\ -e^t & e^{3t} \end{bmatrix}$, $\qquad x(t) = \dfrac{1}{2}\begin{bmatrix} 5e^t + e^{3t} \\ -5e^t + e^{3t} \end{bmatrix}$

16. $x = \begin{bmatrix} 1 & e^{4t} \\ 2 & -2e^{4t} \end{bmatrix}\begin{bmatrix} 1 & 1 \\ 2 & -2 \end{bmatrix}^{-1}\begin{bmatrix} 2 \\ -1 \end{bmatrix}$

$\quad = \begin{bmatrix} 1 & e^{4t} \\ 2 & -2e^{4t} \end{bmatrix}\begin{bmatrix} 1/2 & 1/4 \\ 1/2 & -1/4 \end{bmatrix}\begin{bmatrix} 2 \\ -1 \end{bmatrix} = \begin{bmatrix} (3 + 5e^{4t})/4 \\ (6 - 10e^{4t})/4 \end{bmatrix}$

17. $\varphi(t) = \begin{bmatrix} 5\cos 4t & -5\sin 4t \\ 2\cos 4t + 4\sin 4t & 4\cos 4t - 2\sin 4t \end{bmatrix}$,

$\quad x(t) = \dfrac{1}{2}\begin{bmatrix} -5\sin 4t \\ 4\cos 4t - 2\sin 4t \end{bmatrix}$

18. $x = e^{2t}\begin{bmatrix} 1 & 2+t \\ 1 & 1+t \end{bmatrix}\begin{bmatrix} 1 & 2 \\ 1 & 1 \end{bmatrix}^{-1}\begin{bmatrix} 1 \\ 0 \end{bmatrix}$

$\quad x = e^{2t}\begin{bmatrix} 1 & 2+t \\ 1 & 1+t \end{bmatrix}\begin{bmatrix} -1 & 2 \\ 1 & -1 \end{bmatrix}\begin{bmatrix} 1 \\ 0 \end{bmatrix} = e^{2t}\begin{bmatrix} 1+t \\ t \end{bmatrix}$

19. $\varphi(t) = \begin{bmatrix} 2 \cos 3t & -2 \sin 3t \\ -3 \cos 3t + 3 \sin 3t & 3 \cos 3t + 3 \sin 3t \end{bmatrix}$,

$x(t) = \dfrac{1}{3} \begin{bmatrix} 3 \cos 3t - \sin 3t \\ -3 \cos 3t + 6 \sin 3t \end{bmatrix}$

20. $\Phi(t) = e^{5t} \begin{bmatrix} 5 \cos 4t & 5 \sin 4t \\ 2 \cos 4t + 4 \sin 4t & 2 \sin 4t - 4 \cos 4t \end{bmatrix}$

$\Phi(0) = \begin{bmatrix} 5 & 0 \\ 2 & -4 \end{bmatrix} \qquad \Phi(0)^{-1} = \begin{bmatrix} 1/5 & 0 \\ 1/10 & -1/4 \end{bmatrix}$

$x = \Phi(t) \ \Phi(0)^{-1} \begin{bmatrix} 2 \\ 0 \end{bmatrix} = \Phi(t) \begin{bmatrix} 2/5 \\ 1/5 \end{bmatrix}$

$= e^{5t} \begin{bmatrix} 2 \cos 4t + \sin 4t \\ 2 \sin 4t \end{bmatrix}$

21. $\Phi(t) = \begin{bmatrix} 6 & 3e^t & 2e^{-t} \\ 2 & e^t & e^{-t} \\ 5 & 2e^t & 2e^{-t} \end{bmatrix}$

$x(t) = \begin{bmatrix} -12 + 12e^t + 2e^{-t} \\ -4 + 4e^t + e^{-t} \\ -10 + 8e^t + 2e^{-t} \end{bmatrix}$

22. $\Phi(t) = \begin{bmatrix} e^{3t} & e^t & 0 \\ -e^{3t} & -e^t & e^{-t} \\ e^{3t} & 0 & -e^{-t} \end{bmatrix}$

$$\Phi(0) = \begin{bmatrix} 1 & 1 & 0 \\ -1 & -1 & 1 \\ 1 & 0 & -1 \end{bmatrix}, \qquad \Phi(0)^{-1} = \begin{bmatrix} 1 & 1 & 1 \\ 0 & -1 & -1 \\ 1 & 1 & 0 \end{bmatrix}$$

$$x = \Phi(t) \, \Phi(0)^{-1} \begin{bmatrix} 1 \\ 0 \\ -1 \end{bmatrix} = \Phi(t) \begin{bmatrix} 0 \\ 1 \\ 1 \end{bmatrix} = \begin{bmatrix} e^t \\ e^{-t} - e^t \\ -e^{-t} \end{bmatrix}$$

Each of the systems in Problems 23-32 is of the form

$$x' = Ax + f(t),$$

and a particular solution $x_p(t)$ is given in terms of a fundamental matrix $\Phi(t)$ by the variation of parameters formula

$$x_p(t) = \Phi(t) \int_a^t \Phi(s)^{-1} f(s) \, ds.$$

In each even-numbered problem we give both the fundamental matrix $\Phi(t)$ and the particular solution $x_p(t)$. We take $a = 0$ unless otherwise specified.

23. $x(t) = \dfrac{1}{150} \begin{bmatrix} 666 - 120t - 575t^{-t} - 91e^{5t} \\ 588 - 60t - 575e^{-t} - 13e^{5t} \end{bmatrix}$

24. $\Phi(t) = \begin{bmatrix} e^{-3t} & 2e^{2t} \\ -2e^{-3t} & e^{2t} \end{bmatrix}$

$$x_p(t) = \Phi(t) \int_0^t \frac{1}{5} e^s \begin{bmatrix} e^{2s} & -2e^{2s} \\ 2e^{-3s} & e^{-3s} \end{bmatrix} \begin{bmatrix} 3e^{2s} \\ 5s \end{bmatrix} ds$$

$$= \frac{1}{900} \begin{bmatrix} -250 - 1500t + 558e^{2t} - 308e^{-3t} + 2160te^{2t} \\ -625 + 750t + 9e^{2t} + 616e^{-3t} + 1080te^{2t} \end{bmatrix}$$

25. $x(t) = \dfrac{1}{16} \begin{bmatrix} 3e^{3t} - 3e^{-t} + 4te^{-t} \\ 3e^{3t} - 3e^{-t} + 20te^{-t} \end{bmatrix}$

26. $\Phi(t) = \begin{bmatrix} 1+t & 1 \\ 2+3t & 3 \end{bmatrix}$

$$x_p(t) = \Phi(t) \int_1^t \begin{bmatrix} 3 & -1 \\ -2-3s & 1+s \end{bmatrix} \begin{bmatrix} 1/s^2 \\ 1/s^3 \end{bmatrix} ds$$

$$= \frac{1}{2t^2} \begin{bmatrix} 5t^3 - t^2 - 3t - 6t^2 \ln t \\ 15t^3 - 8t^2 - 3t - 1 - 18t^2 \ln t \end{bmatrix}$$

27. $x(t) = \dfrac{1}{28} \begin{bmatrix} -6 \cos 4t + 28 \sin 4t + 8 \cos 3t - 28 \sin 3t \\ -31 \cos 4t + 8 \sin 4t + 32 \cos 3t - 8 \sin 3t \end{bmatrix}$

28. $\Phi(t) = \begin{bmatrix} 5\sin 4t & 5\cos 4t \\ 3\sin 4t - 4\cos 4t & 3\cos 4t + 4\sin 4t \end{bmatrix}$

$$x_p(t) = \frac{1}{20} \Phi(t) \int_0^t \begin{bmatrix} 3\cos 4s + 4\sin 4s & -5\cos 4s \\ 4\cos 4s - 3\sin 4s & 5\sin 4s \end{bmatrix} \begin{bmatrix} \cos 4s \\ \sin 4s \end{bmatrix} ds$$

$$= \frac{1}{320} \Phi(t) \begin{bmatrix} 24t + 3\sin 8t - 2\sin^2 4t \\ 72t - \sin 8t - 6\sin^2 4t \end{bmatrix}$$

29. $x(t) = \Phi(t) \begin{bmatrix} t - \sin t + \ln|\cos t + \sin t \cos t| \\ 1 - 2t - \cos t + \ln|\cos t| \end{bmatrix}$

where $\Phi(t) = \begin{bmatrix} \cos t & -\sin t \\ 2 \cos t + \sin t & \cos t - 2 \sin t \end{bmatrix}$

30. $\Phi(t) = e^{2t} \begin{bmatrix} \cos t & -\sin t \\ \sin t & \cos t \end{bmatrix}$

$x_p(t) = \Phi(t) \int_0^t e^{-2s} \begin{bmatrix} \cos s & \sin s \\ -\sin s & \cos s \end{bmatrix} \begin{bmatrix} e^{2s}\tan s \\ 0 \end{bmatrix} ds$

$= \Phi(t) \begin{bmatrix} 1 - \cos t \\ \sin t - \ln|\sec t + \tan t| \end{bmatrix}$

31. $x(t) = \frac{1}{4} e^{3t} \begin{bmatrix} -2t \sin 2t \\ \sin 2t + 2t \cos 2t \end{bmatrix}$

32. $\Phi(t) = \begin{bmatrix} 3+2t & 2 \\ 1+t & 1 \end{bmatrix}$

$x_p(t) = \Phi(t) \int_1^t \begin{bmatrix} 1 & -2 \\ -1-s & 3+2s \end{bmatrix} \begin{bmatrix} \ln s \\ s \end{bmatrix} ds$

$= \frac{1}{12} \Phi(t) \begin{bmatrix} 24 -12t -12t^2+ 12t \ln t \\ -41 +12t +21t^2+ 8t^3- 12t \ln t - 6t^2\ln t \end{bmatrix}$

MATRIX EXPONENTIALS AND LINEAR SYSTEMS

In each of Problems 1-12 we give the eigenvector matrix P, the diagonal eigenvalue matrix D, and the fundamental matrix $\Phi(t) = Pe^{Dt}P^{-1}$.

1. $P = \begin{bmatrix} 2 & 1 \\ 1 & 1 \end{bmatrix}$ $\qquad$ $D = \begin{bmatrix} 3 & 0 \\ 0 & 1 \end{bmatrix}$

$\qquad \Phi(t) = \begin{bmatrix} 2e^{3t} - e^t & -2e^{3t} + 2e^t \\ e^{3t} - e^t & -e^{3t} + 2e^t \end{bmatrix}$

2. $P = \begin{bmatrix} 3 & 1 \\ 2 & 1 \end{bmatrix}$ $\qquad$ $D = \begin{bmatrix} 2 & 0 \\ 0 & 0 \end{bmatrix}$

$\qquad \Phi(t) = \begin{bmatrix} 3e^{2t} - 2 & -3e^{2t} + 3 \\ 2e^{2t} - 2 & -2e^{2t} + 3 \end{bmatrix}$

3. $P = \begin{bmatrix} 3 & 1 \\ 2 & 1 \end{bmatrix}$ $\qquad$ $D = \begin{bmatrix} 3 & 0 \\ 0 & 2 \end{bmatrix}$

$\qquad \Phi(t) = \begin{bmatrix} 3e^{3t} - 2e^{2t} & -3e^{3t} + 3e^{2t} \\ 2e^{3t} - 2e^{2t} & -2e^{3t} + 3e^{2t} \end{bmatrix}$

4. $P = \begin{bmatrix} 4 & 1 \\ 3 & 1 \end{bmatrix}$ $\qquad$ $D = \begin{bmatrix} 2 & 0 \\ 0 & 1 \end{bmatrix}$

$$\Phi(t) = \begin{bmatrix} 4e^{2t} - 3e^t & - 4e^{2t} + 4e^t \\ 3e^{2t} - 3e^t & - 3e^{2t} + 4e^t \end{bmatrix}$$

5. $P = \begin{bmatrix} 4 & 1 \\ 3 & 1 \end{bmatrix}$ $D = \begin{bmatrix} 3 & 0 \\ 0 & 1 \end{bmatrix}$

$$\Phi(t) = \begin{bmatrix} 4e^{3t} - 3e^t & - 4e^{3t} + 4e^t \\ 3e^{3t} - 3e^t & - 3e^{3t} + 4e^t \end{bmatrix}$$

6. $P = \begin{bmatrix} 3 & 2 \\ 4 & 3 \end{bmatrix}$ $D = \begin{bmatrix} 2 & 0 \\ 0 & 1 \end{bmatrix}$

$$\Phi(t) = \begin{bmatrix} 9e^{2t} - 8e^t & - 6e^{2t} + 6e^t \\ 12e^{2t} - 12e^t & - 8e^{2t} + 9e^t \end{bmatrix}$$

7. $P = \begin{bmatrix} 5 & 2 \\ 2 & 1 \end{bmatrix}$ $D = \begin{bmatrix} 2 & 0 \\ 0 & 1 \end{bmatrix}$

$$\Phi(t) = \begin{bmatrix} 5e^{2t} - 4e^t & -10e^{2t} + 10e^t \\ 2e^{2t} - 2e^t & - 4e^{2t} + 5e^t \end{bmatrix}$$

8. $P = \begin{bmatrix} 5 & 3 \\ 3 & 2 \end{bmatrix}$ $D = \begin{bmatrix} 2 & 0 \\ 0 & 1 \end{bmatrix}$

$$\Phi(t) = \begin{bmatrix} 10e^{2t} - 9e^t & -15e^{2t} + 15e^t \\ 6e^{2t} - 6e^t & -9e^{2t} + 10e^t \end{bmatrix}$$

9.　$P = \begin{bmatrix} 1 & 1 \\ 1 & -1 \end{bmatrix}$　　$D = \begin{bmatrix} 4 & 0 \\ 0 & 2 \end{bmatrix}$

$$\Phi(t) = \frac{1}{2} \begin{bmatrix} e^{4t} + e^{2t} & e^{4t} - e^{2t} \\ e^{4t} - e^{2t} & e^{4t} + e^{2t} \end{bmatrix}$$

10.　$P = \begin{bmatrix} 1 & 1 \\ 1 & -1 \end{bmatrix}$　　$D = \begin{bmatrix} 6 & 0 \\ 0 & 2 \end{bmatrix}$

$$\Phi(t) = \frac{1}{2} \begin{bmatrix} e^{6t} + e^{2t} & e^{6t} - e^{2t} \\ e^{6t} - e^{2t} & e^{6t} + e^{2t} \end{bmatrix}$$

11.　$P = \begin{bmatrix} 2 & -1 \\ 1 & 2 \end{bmatrix}$　　$D = \begin{bmatrix} 10 & 0 \\ 0 & 5 \end{bmatrix}$

$$\Phi(t) = \frac{1}{5} \begin{bmatrix} 4e^{10t} + e^{5t} & 2e^{10t} - 2e^{5t} \\ 2e^{10t} - 2e^{5t} & e^{10t} + 4e^{5t} \end{bmatrix}$$

12.　$P = \begin{bmatrix} 2 & -1 \\ 1 & 2 \end{bmatrix}$　　$D = \begin{bmatrix} 15 & 0 \\ 0 & 5 \end{bmatrix}$

$$\Phi(t) = \frac{1}{5} \begin{bmatrix} 4e^{15t} + e^{5t} & 2e^{15t} - 2e^{5t} \\ 2e^{15t} - 2e^{5t} & e^{15t} + 4e^{5t} \end{bmatrix}$$

13. $A = I + B$ where $B^2 = 0$, so

$$e^{at} = e^{It}e^{bt} = e^{It}(I + Bt)$$

$$= e^t \begin{bmatrix} 1 & 2t \\ 0 & 1 \end{bmatrix}$$

14. $A = 3I + B$ where $B^2 = 0$, so

$$e^{at} = e^{3It}e^{bt} = e^{3It}(I + Bt)$$

$$= e^{3t} \begin{bmatrix} 1 & 5t \\ 0 & 1 \end{bmatrix}$$

15. $A = I + B$ where

$$B^2 = \begin{bmatrix} 0 & 0 & 1 \\ 0 & 0 & 0 \\ 0 & 0 & 0 \end{bmatrix}$$

and $B^3 = 0$. Therefore

$$e^{at} = e^{It}e^{bt} = e^{It}(I + Bt + B^2t^2/2)$$

$$= e^t \begin{bmatrix} 1 & t & t^2/2 \\ 0 & 1 & t \\ 0 & 0 & 1 \end{bmatrix}$$

16.   A = 2I + B   where

$$B^2 = \begin{bmatrix} 0 & 0 & 9 \\ 0 & 0 & 0 \\ 0 & 0 & 0 \end{bmatrix}$$

and $B^3 = 0$.   Therefore

$$e^{at} = e^{2it}e^{bt} = e^{2it}(I + Bt + B^2t^2/2)$$

$$= e^{2t} \begin{bmatrix} 1 & 3t & 13t^2/2 \\ 0 & 1 & 3t \\ 0 & 0 & 1 \end{bmatrix}$$

In Problems 17-20 we compute the particular solution

$$x(t) = e^{At} \int e^{-At}f(t)\ dt$$

as in Example 4.   In Problems 17 and 18 we use the exponential matrix e^{at} calculated in Problem 13 above, while in Problems 19 and 20 we use the matrices of Problems 15 and 16, respectively.

17. $$x(t) = e^t \begin{bmatrix} 1 & 2t \\ 0 & 1 \end{bmatrix} \int e^{-t} \begin{bmatrix} 1 & -2t \\ 0 & 1 \end{bmatrix} \begin{bmatrix} 12e^t \\ 6e^t \end{bmatrix} dt$$

$$= e^t \begin{bmatrix} 1 & 2t \\ 0 & 1 \end{bmatrix} \int \begin{bmatrix} 12 & -12t \\ 6 \end{bmatrix} dt$$

$$x(t) = e^t \begin{bmatrix} 12t + 6t^2 \\ 6t \end{bmatrix}$$

18. $x(t) = e^t \begin{bmatrix} 1 & 2t \\ 0 & 1 \end{bmatrix} \int \begin{bmatrix} 0 \\ 6te^{-t} \end{bmatrix} dt$

$= e^t \begin{bmatrix} 1 & 2t \\ 0 & 1 \end{bmatrix} \begin{bmatrix} 0 \\ -6(t+1)e^{-t} \end{bmatrix} = \begin{bmatrix} -12t-12t^2 \\ -6-6t \end{bmatrix}$

19. $\int e^{-t} \begin{bmatrix} 1 & -t & t^2/2 \\ 0 & 1 & -t \\ 0 & 0 & 1 \end{bmatrix} \begin{bmatrix} 6 \\ 12 \\ 24 \end{bmatrix} te^t \, dt = \begin{bmatrix} 3t^2-4t^3+3t^4 \\ 6t^2-8t^3 \\ 12t^2 \end{bmatrix}$

so $x(t) = e^t \begin{bmatrix} 1 & t & t^2/2 \\ 0 & 1 & t \\ 0 & 0 & 1 \end{bmatrix} \begin{bmatrix} 3t^2-4t^3+3t^4 \\ 6t^2-8t^3 \\ 12t^2 \end{bmatrix}$

$= e^t \begin{bmatrix} 3t^2 + 2t^3 + t^4 \\ 6t^2 + 4t^3 \\ 12t^2 \end{bmatrix}$

20. $\int e^{-2t} \begin{bmatrix} 1 & -3t & 13t^2/2 \\ 0 & 1 & -3t \\ 0 & 0 & 1 \end{bmatrix} \begin{bmatrix} 8t^3 \\ 6t^2 \\ 4t \end{bmatrix} e^{2t} \, dt = \begin{bmatrix} 4t^4 \\ -2t^3 \\ 2t^2 \end{bmatrix}$

so $x(t) = e^{2t} \begin{bmatrix} 1 & 3t & 13t^2/2 \\ 0 & 1 & 3t \\ 0 & 0 & 1 \end{bmatrix} \begin{bmatrix} 4t^4 \\ -2t^3 \\ 2t^2 \end{bmatrix} = e^{2t} \begin{bmatrix} 11t^4 \\ 4t^3 \\ 2t^2 \end{bmatrix}$

23. $e^{At} = I \cosh t + A \sinh t = \begin{bmatrix} \cosh t & \sinh t \\ \sinh t & \cosh t \end{bmatrix}$

so the general solution of $x' = Ax$ is

$$x(t) = e^{At}c = \begin{bmatrix} c_1 \cosh t + c_2 \sinh t \\ c_1 \sinh t + c_2 \cosh t \end{bmatrix}$$

24. Direct calculation gives $A^2 = -4I$, and it follows that

$$A^3 = -4A \quad \text{and} \quad A^4 = 16I.$$

Therefore

$$e^{at} = I + At - 4It^2/2! - 4At^3/3!$$

$$+ 16It^4/4! + 16At^5/5! \cdots$$

$$= I [1 - (2t)^2/2! + (2t)^4/4! - \cdots]$$

$$+ (A/2)[(2t) - (2t)^3/3! + (2t)^5/5! - \cdots]$$

$$e^{at} = I \cos 2t + (A/2) \sin 2t.$$

25. $A = I + B$ where

$$B = \begin{bmatrix} 0 & 2 & 0 \\ 0 & 0 & 2 \\ 0 & 0 & 0 \end{bmatrix}, \qquad B^2 = \begin{bmatrix} 0 & 0 & 4 \\ 0 & 0 & 0 \\ 0 & 0 & 0 \end{bmatrix}$$

and $B^3 = 0$. Therefore the general solution is

$$x(t) = e^{At}c = e^{It}e^{Bt}c$$

$$= e^t [I + Bt + B^2t^2/2]c$$

$$= e^t \begin{bmatrix} 1 & 2t & 2t^2 \\ 0 & 1 & 2t \\ 0 & 0 & 1 \end{bmatrix} \begin{bmatrix} c_1 \\ c_2 \\ c_3 \end{bmatrix}$$

$$x(t) = e^t \begin{bmatrix} c_1 + 2c_2t + 2c_3t^2 \\ c_2 + 2c_3t \\ c_3 \end{bmatrix}.$$

CHAPTER 6

NUMERICAL METHODS

The typical problem in Chapter 6 calls for a table of approximate
values. Each of these tables was produced on a microcomputer (an
IBM Personal Computer) using programs similar to those listed in
the text. However, to save space in this brief manual, we have
chosen in most cases to give only data from the last line of the
table -- that is, only the values at the final point of the
interval in question. This will enable the reader to determine
whether his or her results agree with ours (though answers
produced with different hardware and/or software can be expected
to differ slightly -- perhaps in the last digit or two -- because
of differing methods of performing simple arithmetic).

SECTION 6.1

INTRODUCTION: EULER'S METHOD

1.   The iterative formula of Euler's method is

$$y_{n+1} = y_n + h(-y_n),$$

and the exact solution is $y(x) = 2e^{-x}$. The resulting table
of approximate and actual values is

x	Y with h = 0.1	Y with h = 0.05	Y exact
0.0	2.0000	2.0000	2.0000
0.1	1.8000	1.8050	1.8097
0.2	1.6200	1.6290	1.6375
0.3	1.4580	1.4702	1.4816
0.4	1.3122	1.3268	1.3406
0.5	1.1810	1.1975	1.2131

2. Iterative formula: $y_{n+1} = y_n + h(2y_n)$

   Exact solution: $y(x) = (1/2)e^{2x}$

   Approx: $y(0.5) \approx 1.2442$ with $h = 0.1$,

   1.2969 with $h = 0.05$

   Actual: $y(0.5) \approx 1.3591$

3. Iterative formula: $y_{n+1} = y_n + h(y_n + 1)$

   Exact solution: $y(x) = 2e^x - 1$

   Approx: $y(0.5) \approx 2.2210$ with $h = 0.1$,

   2.2578 with $h = 0.05$

   Actual: $y(0.5) \approx 2.2974$

4. Iterative formula: $y_{n+1} = y_n + h(x_n - y_n)$

   Exact solution: $y(x) = 2e^{-x} + x - 1$

   Approx: $y(0.5) \approx 0.6810$ with $h = 0.1$,

   0.6975 with $h = 0.05$

   Actual: $y(0.5) \approx 0.7131$

5. Iterative formula: $y_{n+1} = y_n + h(y_n - x_n - 1)$

   Exact solution: $y(x) = 2 + x - e^x$

   Approx: $y(0.5) \approx 0.8895$ with $h = 0.1$,

   0.8711 with $h = 0.05$

   Actual: $y(0.5) \approx 0.8513$

6. Iterative formula: $y_{n+1} = y_n + h(-2x_ny_n)$

   Exact solution: $y(x) = 2 \exp(-x^2)$

   Approx: $y(0.5) \approx 1.6272$ with $h = 0.1$,

   1.5912 with $h = 0.05$

   Actual: $y(0.5) \approx 1.5576$

7. Iterative formula: $y_{n+1} = y_n + h(-3x_n^2 y_n)$

Exact solution: $y(x) = 3 \exp(-x^3)$

Approx: $y(0.5) \approx 2.7373$ with $h = 0.1$,

2.6930 with $h = 0.05$

Actual: $y(0.5) \approx 2.6475$

8. Iterative formula: $y_{n+1} = y_n + h \exp(-y_n)$

Exact solution: $y(x) = \ln(x + 1)$

Approx: $y(0.5) \approx 0.4198$ with $h = 0.1$,

0.4124 with $h = 0.05$

Actual: $y(0.5) \approx 0.4055$

9. Iterative formula: $y_{n+1} = y_n + h(1 + y_n^2)/4$

Exact solution: $y(x) = \tan[(x + \pi)/4]$

Approx: $y(0.5) \approx 1.2785$ with $h = 0.1$,

1.2828 with $h = 0.05$

Actual: $y(0.5) \approx 2.2874$

10. Iterative formula: $y_{n+1} = y_n + h(2x_n y_n^2)$

Exact solution: $y(x) = 1/(1 - x^2)$

Approx: $y(0.5) \approx 1.2313$ with $h = 0.1$,

1.2776 with $h = 0.05$

Actual: $y(0.5) \approx 1.3333$

The tables of approximate and actual values called for in Problems 11-16 were produced using a slight alteration of Program EULERANS, which is listed below preceding Problem 17.

11. The iterative formula of Euler's method is

$$y_{n+1} = y_n + h(y_n - 2),$$

and the exact solution is $y(x) = 2 - e^x$. The resulting table of approximate and actual values is

x	Y with h = 0.01	Y with h = 0.005	Y actual
0.0	1.0000	1.0000	1.0000
0.2	0.7798	0.7792	0.7786
0.4	0.5111	0.5097	0.5082
0.6	0.1833	0.1806	0.1779
0.8	-0.2167	-0.2211	-0.2255
1.0	-0.7048	-0.7115	-0.7183

12. Iterative formula: $y_{n+1} = y_n + h(y_n - 1)^2/2$

Exact solution: $y(x) = 1 + 2/(2 - x)$

Approx: $y(1) \approx 2.9864$ with $h = 0.01$,

2.9931 with $h = 0.005$

Actual: $y(1) \approx 3.0000$

13. Iterative formula: $y_{n+1} = y_n + 2hx_n^3/y_n$

Exact solution: $y(x) = (8 + x^4)^{1/2}$

Approx: $y(2) \approx 4.8890$ with $h = 0.01$,

4.8940 with $h = 0.005$

Actual: $y(2) \approx 4.8990$

14. Iterative formula: $y_{n+1} = y_n + hy_n^2/x_n$

Exact solution: $y(x) = 1/(1 - \ln x)$

Approx: $y(2) \approx 3.2031$ with $h = 0.01$,

3.2304 with $h = 0.005$

Actual: $y(2) \approx 3.2589$

15. Iterative formula: $y_{n+1} = y_n + h(3 - 2y_n/x_n)$

Exact solution: $y(x) = x + 4/x^2$

Approx: $y(3) \approx 3.4422$ with $h = 0.01$,

3.4433 with $h = 0.005$

Actual: $y(3) \approx 3.4444$

16. Iterative formula: $y_{n+1} = y_n + 2hx_n^5/y_n^2$

Exact solution: $y(x) = (x^6 - 37)^{1/3}$

Approx: $y(3) \approx 8.8440$ with $h = 0.01$,

8.8445 with $h = 0.005$

Actual: $y(3) \approx 8.8451$

The tables of approximate values called for in Problems 17-24 were produced using the following program, which upon execution first asks for the righthand-side in the differential equation $y' = f(x,y)$ to be edited into line 230.

```
100 REM--Program EULERANS
110 REM--Uses Euler's method to approximate the solution
120 REM--of the equation y'=f(x,y) on the given interval.
130 REM--The function f(x,y) is edited into line 230, and
140 REM--and the number N of subintervals is input.
150 REM
160 REM--Initialization:
165 REM
170       PRINT "EDIT IN YOUR FUNCTION, THEN RUN 210"
180       PRINT :   EDIT 230
210       DEFDBL A,F,H,X,Y   :   DEFINT I,J,K,N,S
220       DIM A(4,20)
230       DEF FNF(X,Y) = X*X + Y*Y
245       INPUT "INITIAL X"; X0
250       INPUT "INITIAL Y"; Y0
260       INPUT "INITIAL NUMBER OF SUBINTERVALS"; N
270       K = 1   :   H = .1
280       PRINT "    Y with   Y with   Y with    Y with"
290       PRINT " X   h=0.1    h=0.02   h=0.004  h=0.0008"
300       PRINT
310 REM
320 REM--Euler's iteration:
325 REM
```

```
330        FOR I = 1 TO 4
340            X = X0   :    Y = Y0
350            FOR J = 0 TO N
360                IF J/K = J\K THEN A(I,J/K) = Y
370                Y = Y + H*FNF(X,Y)
380                X = X + H
390            NEXT J
400            H = H/5  :  K = 5*K  :  N = 5*N
410        NEXT I
420 REM
425 REM--Print results:
430 REM
435        N = N/625  :   S = N/5
440        FOR J = 0 TO N STEP S
445            X = X0 + J/10
450            PRINT USING
               "#.#   #.####   #.####   #.####   #.####";
               X,A(1,J), A(2,J),  A(3,J), A(4,J)
460        NEXT J
470        END
```

17. Here $f(x,y) = x^2 + y^2$ on $[0,1]$. A run of Program
EULERANS results in the following display:


```
RUN
EDIT IN YOUR FUNCTION, THEN RUN 210

230        DEF FNF(X,Y) = X*X + Y*Y
RUN 210
INITIAL X? 0
INITIAL Y? 0
INITIAL NUMBER OF SUBINTERVALS? 10
```

	Y with	Y with	Y with	Y with
x	h=0.1	h=0.02	h=0.004	h=0.0008
0.0	0.0000	0.0000	0.0000	0.0000
0.2	0.0010	0.0023	0.0026	0.0027
0.4	0.0140	0.0198	0.0210	0.0213
0.6	0.0551	0.0688	0.0717	0.0723
0.8	0.1413	0.1672	0.1727	0.1738
1.0	0.2925	0.3379	0.3477	0.3497

It seems apparent that $y(1) \approx 0.35$, in contrast with
Example 2 in the text, where the initial condition is $y(0) = 1$.

In Problems 18-24 we give only the final approximate values of y
obtained using Euler's method with step sizes $h = 0.1$, $h = 0.02$, $h = 0.004$, and $h = 0.0008$.

18. With $x_0 = 0$ and $y_0 = 1$, the approximate values of $y(2)$ obtained are:

h	0.1	0.02	0.004	0.0008
y	1.6680	1.6771	1.6790	1.6794

19. With $x_0 = 0$ and $y_0 = 1$, the approximate values of $y(2)$ obtained are:

h	0.1	0.02	0.004	0.0008
y	6.1831	6.3653	6.4022	6.4096

20. With $x_0 = 0$ and $y_0 = -1$, the approximate values of $y(2)$ obtained are:

h	0.1	0.02	0.004	0.0008
y	-1.3792	-1.2843	-1.2649	-1.2610

21. With $x_0 = 1$ and $y_0 = 2$, the approximate values of $y(2)$ obtained are:

h	0.1	0.02	0.004	0.0008
y	2.8508	2.8681	2.8716	2.8723

22. With $x_0 = 0$ and $y_0 = 1$, the approximate values of $y(2)$ obtained are:

h	0.1	0.02	0.004	0.0008
y	6.9879	7.2601	7.3154	7.3264

23. With $x_0 = 0$ and $y_0 = 0$, the approximate values of $y(1)$ obtained are:

h	0.1	0.02	0.004	0.0008
y	1.2262	1.2300	1.2306	1.2307

24. With $x_0 = -1$ and $y_0 = 1$, the approximate values of $y(1)$ obtained are:

h	0.1	0.02	0.004	0.0008
y	0.9585	0.9918	0.9984	0.9997

25. With step sizes $h = 0.15$, $h = 0.03$, and $h = 0.006$ we get the following results:

X	Y with h=0.15	Y with h=0.03	Y with h=0.006
-1.0	1.0000	1.0000	1.0000
-0.7	1.0472	1.0512	1.0521
-0.4	1.1213	1.1358	1.1390
-0.1	1.2826	1.3612	1.3835
+0.2	0.8900	1.4711	0.8210
+0.5	0.7460	1.2808	0.7192

While the values for $h = 0.15$ alone are not conclusive, a comparison of the values of y for all three step sizes with $x > 0$ suggests some anomaly in the transition from negative to positive values of x.

26. With step sizes $h = 0.1$ and $h = 0.01$ we get the following results:

X	Y with h = 0.1	Y with h = 0.01
0.0	0.0000	0.0000
0.1	0.0000	0.0003
0.2	0.0010	0.0025
0.3	0.0050	0.0086
.	.	.
.	.	.
.	.	.

1.8	2.8200	4.3308
1.9	3.9393	7.9425
2.0	5.8521	28.3926

Clearly there is some difficulty near   x = 2.

27. With step sizes   h = 0.1   and   h = 0.01   we get the
     following results:

X	Y with h = 0.1	Y with h = 0.01
0.0	1.0000	1.0000
0.1	1.2000	1.2200
0.2	1.4428	1.4967
.	.	.
.	.	.
.	.	.
0.7	4.3460	6.4643
0.8	5.8670	11.8425
0.9	8.3349	39.5010

Clearly there is some difficulty near   x = 0.9.

SECTION 6.2

A CLOSER LOOK AT THE EULER METHOD, AND IMPROVEMENTS

The initial value problems in Problems 1-24 here are the same as
those in Problems 1-24 of Section 6.1, so the answers here may be
compared with those in the previous section, to see how great an
improvement is the improved Euler method over the original Euler
method.

Problems 1-5 involve linear equations of the form

     y' = ax + by + c.

For such an equation it is readily verified that the iterative

formulas of the improved Euler method and the 3-term Taylor
series method both reduce to the same formula,

$$y_{n+1} = y_n + h(ax_n + by_n + c) + (h^2/2)[a + b(ax_n + by_n + c)].$$

Hence in this case the two methods give identical results.

In each of Problems 1-10 we give first the predictor formula for
u_{n+1}, next the improved Euler corrector for y_{n+1}, then the
3-term Taylor series formula for y_{n+1}, and finally the
approximate and actual values of $y(0.5)$ obtained with step size
$h = 0.1$.

1. $u_{n+1} = y_n - hy_n$

   $y_{n+1} = y_n + (h/2)[- y_n - u_{n+1}]$

   $y_{n+1} = y_n - hy_n + h^2y_n/2$

   Approx value: 1.2141

   Actual value: 1.2131

   The complete table of results is

	X	Improved Euler Y	Three-term Taylor Y	Actual value
	0.1	1.8100	1.8100	1.8097
	0.2	1.6381	1.6381	1.6375
	0.3	1.4824	1.4824	1.4816
	0.4	1.3416	1.3416	1.3406
	0.5	1.2142	1.2142	1.2131^O

2. $u_{n+1} = y_n + 2hy_n$

   $y_{n+1} = y_n + (h/2)[2y_n + 2u_{n+1}]$

   $y_{n+1} = y_n + 2hy_n + 2h^2y_n$

   Approx value: 1.3514

   Actual value: 1.3591

3. $u_{n+1} = y_n + h(y_n + 1)$

$y_{n+1} = y_n + (h/2)[(y_n + 1) + (u_{n+1} + 1)]$

$y_{n+1} = y_n + (h + h^2/2)(y_n + 1)$

Approx value: 2.2949

Actual value: 2.2974

4. $u_{n+1} = y_n + h(x_n - y_n)$

$y_{n+1} = y_n + (h/2)[(x_n - y_n) + (x_n + h - u_{n+1})]$

$y_{n+1} = y_n + h(x_n - y_n) + h^2(1 - x_n + y_n)/2$

Approx value: 0.7142

Actual value: 0.7131

5. $u_{n+1} = y_n + h(y_n - x_n - 1)$

$y_{n+1} = y_n + (h/2)[(y_n - x_n - 1) + (u_{n+1} - x_n - h - 1)]$

$y_{n+1} = y_n + h(y_n - x_n - 1) + h^2(y_n - x_n - 2)/2$

Approx value: 0.8526

Actual value: 0.8513

6. $u_{n+1} = y_n - 2x_n y_n h$

$y_{n+1} = y_n - (h/2)[2x_n y_n + 2(x_n + h)u_{n+1}]$

$y_{n+1} = y_n - 2x_n y_n h + h^2(2x_n^2 - 1)y_n$

Improved Euler value: 1.5575

3-term Taylor value: 1.5542

Actual value: 1.5576

The complete table of results is

X	Improved Euler Y	Three-term Taylor Y	Actual value
0.1	1.9800	1.9800	1.9801
0.2	1.9214	1.9210	1.9216
0.3	1.8276	1.8265	1.8279
0.4	1.7041	1.7019	1.7043
0.5	1.5575	1.5542	1.5576

7. $u_{n+1} = y_n - 3x_n^2 y_n h$

$y_{n+1} = y_n - (h/2)[3x_n^2 y_n + 3(x_n + h)^2 u_{n+1}]$

$y_{n+1} = y_n - 3x_n^2 y_n h + (9x_n^4 - 6x_n)y_n h^2/2$

Improved Euler value: 2.6405

3-term Taylor value: 2.6580

Actual value: 2.6475

8. $u_{n+1} = y_n + h \exp(-y_n)$

$y_{n+1} = y_n + (h/2)[\exp(-y_n) + \exp(-u_{n+1})]$

$y_{n+1} = y_n + h \exp(-y_n) - (h^2/2)\exp(-y_n)$

Improved Euler value: 0.4053

3-term Taylor value: 0.4046

Actual value: 0.4055

9. $u_{n+1} = y_n + h(1 + y_n^2)/4$

$y_{n+1} = y_n + h[1 + y_n^2 + 1 + (u_{n+1})^2]/8$

$y_{n+1} = y_n + (1 + y_n^2)(h/4 + h^2 y_n/16)$

Improved Euler value: 1.2873

3-term Taylor value: 1.2871

Actual value: 1.2874

10. $u_{n+1} = y_n + 2x_ny_n^2h$

$y_{n+1} = y_n + h[x_ny_n^2 + (x_n + h)(u_{n+1})^2]$

$y_{n+1} = y_n + 2x_ny_n^2h + y_n^2(1 + 4x_n^2y_n)h^2$

Improved Euler value: 1.3309

3-term Taylor value: 1.3233

Actual value: 1.3333

The results given below for Problems 11-24 were computed using an improved Euler alteration of Program EULERANS, which was listed in Section 6.1 of this manual.

11. With h = 0.01: $y(1) \approx -0.71824$
     With h = 0.005: $y(1) \approx -0.71827$
              Actual: $y(1) \approx -0.71828$

The table of numerical results is

X	Y with h=0.01	Y with h=0.005	Y exact
0.0	1.00000	1.00000	1.00000
0.2	0.77860	0.77860	0.77860
0.4	0.50819	0.50818	0.50818
0.6	0.17790	0.17789	0.17788
0.8	-0.22551	-0.22553	-0.22554
1.0	-0.71824	-0.71827	-0.71828

12. With h = 0.01: $y(1) \approx 2.99995$
     With h = 0.005: $y(1) \approx 2.99999$
              Actual: $y(1) \approx 3.00000$

13. With h = 0.01: $y(2) \approx 4.89901$
     With h = 0.005: $y(2) \approx 4.89899$
              Actual: $y(2) \approx 4.89898$

14. With h = 0.01: $y(2) \approx 3.25847$
     With h = 0.005: $y(2) \approx 3.25878$
              Actual: $y(2) \approx 3.25889$

15. With h = 0.01: y(3) ≈ 3.44445
    With h = 0.005: y(3) ≈ 3.44445
            Actual: y(3) ≈ 3.44444

16. With h = 0.01: y(3) ≈ 8.84511
    With h = 0.005: y(3) ≈ 8.84509
            Actual: y(3) ≈ 8.84509

17. With h =   0.1: y(1) ≈ 0.35183
    With h =   0.02: y(1) ≈ 0.35030
    With h = 0.004: y(1) ≈ 0.35023
    With h = 0.0008: y(1) ≈ 0.35023

    The table of numerical results is

X	Y with h=0.1	Y with h=0.02	Y with h=0.004	Y with h=0.0008
0.0	0.00000	0.00000	0.00000	0.00000
0.2	0.00300	0.00268	0.00267	0.00267
0.4	0.02202	0.02139	0.02136	0.02136
0.6	0.07344	0.07249	0.07245	0.07245
0.8	0.17540	0.17413	0.17408	0.17408
1.0	0.35183	0.35030	0.35023	0.35023

18. With h =   0.1: y(2) ≈ 1.68043
    With h =   0.02: y(2) ≈ 1.67949
    With h = 0.004: y(2) ≈ 1.67946
    With h = 0.0008: y(2) ≈ 1.67946

19. With h =   0.1: y(2) ≈ 6.40834
    With h =   0.02: y(2) ≈ 6.41134
    With h = 0.004: y(2) ≈ 6.41147
    With h = 0.0008: y(2) ≈ 6.41147

20. With h =   0.1: y(2) ≈ -1.26092
    With h =   0.02: y(2) ≈ -1.26003
    With h = 0.004: y(2) ≈ -1.25999
    With h = 0.0008: y(2) ≈ -1.25999

21. With h =   0.1: $y(2) \approx 2.87204$
     With h =   0.02: $y(2) \approx 2.87245$
     With h =   0.004: $y(2) \approx 2.87247$
     With h = 0.0008: $y(2) \approx 2.87247$

22. With h =   0.1: $y(2) \approx 7.31578$
     With h =   0.02: $y(2) \approx 7.32841$
     With h =   0.004: $y(2) \approx 7.32916$
     With h = 0.0008: $y(2) \approx 7.32920$

23. With h =   0.1: $y(1) \approx 1.22967$
     With h =   0.02: $y(1) \approx 1.23069$
     With h =   0.004: $y(1) \approx 1.23073$
     With h = 0.0008: $y(1) \approx 1.23073$

24. With h =   0.1: $y(1) \approx 1.00006$
     With h =   0.02: $y(1) \approx 1.00000$

     With h =   0.004: $y(1) \approx 1.00000$
     With h = 0.0008: $y(1) \approx 1.00000$

25. $y' = 1 + y^2$
     $y'' = 2yy' = 2y(1 + y^2)$
     $y^{(3)} = 2y' + 6y^2y' = 2(1 + 3y^2)(1 + y^2)$
     $Y_{n+1} = Y_n + (1 + y_n^2)[h + h^2y_n + h^3(1 + 3y_n^2)/3]$

26. $y' = y + y^2$
     $y'' = (1 + 2y)y' = (1 + 2y)(y + y^2)$
     $y^{(3)} = 2y'(y + y^2) + (1 + 2y)(1 + 2y)y'$
     $\qquad = (y + y^2)(1 + 6y + 6y^2)$
     $Y_{n+1} = Y_n + (Y_n + y_n^2)[h + h^2(1 + 2y_n)/2$
     $\qquad\qquad\qquad\qquad + h^3(1 + 6y_n + 6y_n^2)/6]$

27. $y' = (x + y)^2$
     $y'' = 2(x + y)(1 + y') = 2(x + y) + 2(x + y)^3$

$$y^{(3)} = 2(1 + y') + 6(x + y)^2(1 + y')$$

$$= 2 + 8(x + y)^2 + 6(x + y)^4$$

$$y_{n+1} = y_n + h(x_n + y_n)^2 + h^2[(x_n + y_n) + (x_n + y_n)^3]$$

$$+ (h^3/3)[1 + 4(x_n + y_n)^2 + 3(x_n + y_n)^4]$$

SECTION 6.3

THE RUNGE-KUTTA METHOD

1. The actual solution is $y(x) = 2e^{-x}$. The following table of values is obtained by applying the Runge-Kutta method with step size $h = 0.25$:

x	Approx y	Actual y
0.25	1.55762	2.55760
0.50	1.21309	1.21306

2. Actual solution: $y(x) = (1/2)e^{2x}$

    Approx $y(0.5) \approx 1.35867$

    Actual $y(0.5) \approx 1.35914$

3. Actual solution: $y(x) = 2e^x - 1$

    Approx $y(0.5) \approx 2.29740$

    Actual $y(0.5) \approx 2.29744$

4. Actual solution: $y(x) = 2e^{-x} + x - 1$

    Approx $y(0.5) \approx 0.71309$

    Actual $y(0.5) \approx 0.71306$

5. Actual solution: $y(x) = -e^x + x + 2$

    Approx $y(0.5) \approx 0.85130$

    Actual $y(0.5) \approx 0.85128$

6. Actual solution: $y(x) = 2 \exp(-x^2)$

    Approx $y(0.5) \approx 1.55759$

    Actual $y(0.5) \approx 1.55760$

7. Actual solution: $y(x) = 3 \exp(-x^3)$
    Approx $y(0.5) \approx 2.64745$
    Actual $y(0.5) \approx 2.64749$

8. Actual solution: $y(x) = \ln(x + 1)$
    Approx $y(0.5) \approx 0.40547$
    Actual $y(0.5) \approx 0.40547$

9. Actual solution: $y(x) = \tan[(x + \pi)/4]$
    Approx $y(0.5) \approx 1.28743$
    Actual $y(0.5) \approx 1.28743$

10. Actual solution: $y(x) = 1/(1 - x^2)$
    Approx $y(0.5) \approx 1.33337$
    Actual $y(0.5) \approx 1.33333$

11. Actual solution: $y(x) = 2 - e^{-x}$
    With $h = 0.2$: $y(1) \approx -0.71825$
    With $h = 0.1$: $y(1) \approx -0.71828$
            Actual: $y(1) \approx -0.71828$

    The table of numerical results is

X	Y with h=0.2	Y with h=0.1	Y exact
0.0	1.00000	1.00000	1.00000
0.2	0.77860	0.77860	0.77860
0.4	0.50818	0.50818	0.50818
0.6	0.17789	0.17788	0.17788
0.8	-0.22552	-0.22554	-0.22554
1.0	-0.71825	-0.71828	-0.71828

12. Actual solution: $y(x) = 1 + 2/(2 - x)$
    With $h = 0.2$: $y(1) \approx 2.99996$
    With $h = 0.1$: $y(1) \approx 3.00000$
            Actual: $y(1) \approx 3.00000$

13. Actual solution: $y(x) = (8 + x^4)^{1/2}$
    With $h = 0.2$: $y(2) \approx 4.89900$
    With $h = 0.1$: $y(2) \approx 4.89898$
            Actual: $y(2) \approx 4.89898$

14. Actual solution: $y(x) = 1/(1 - \ln x)$

    With h = 0.2: $y(2) \approx 3.25795$

    With h = 0.1: $y(2) \approx 3.25882$

              Actual: $y(2) \approx 3.25889$

15. Actual solution: $y(x) = x + 4/x^2$

    With h = 0.2: $y(3) \approx 3.44445$

    With h = 0.1: $y(3) \approx 3.44444$

              Actual: $y(3) \approx 3.44444$

16. Actual solution: $y(x) = (x^6 - 37)^{1/3}$

    With h = 0.2: $y(3) \approx 8.84515$

    With h = 0.1: $y(3) \approx 8.84509$

              Actual: $y(3) \approx 8.84509$

17. With h =   0.2: $y(1) \approx 0.350258$

    With h =   0.1: $y(1) \approx 0.350234$

    With h =   0.05: $y(1) \approx 0.350232$

    With h = 0.025: $y(1) \approx 0.350232$

    The table of numerical results is

X	Y with h=0.2	Y with h=0.1	Y with h=0.05	Y with h=0.025
0.0	0.000000	0.000000	0.000000	0.000000
0.2	0.002667	0.002667	0.002667	0.002667
0.4	0.021360	0.021359	0.021359	0.021359
0.6	0.072451	0.072448	0.072448	0.072448
0.8	0.174090	0.174081	0.174080	0.174080
1.0	0.350258	0.350234	0.350232	0.350232

18. With h =   0.2: $y(2) \approx 1.679513$

    With h =   0.1: $y(2) \approx 1.679461$

    With h =   0.05: $y(2) \approx 1.679459$

    With h = 0.025: $y(2) \approx 1.679459$

19. With h =   0.2: $y(2) \approx 6.411464$

    With h =   0.1: $y(2) \approx 6.411474$

    With h =   0.05: $y(2) \approx 6.411474$

    With h = 0.025: $y(2) \approx 6.411474$

20. With h =   0.2:   y(2) ≈ -1.259990
     With h =   0.1:   y(2) ≈ -1.259992
     With h =   0.05: y(2) ≈ -1.259993
     With h = 0.025:   y(2) ≈ -1.259993

21. With h =   0.2:   y(2) ≈ 2.872467
     With h =   0.1:   y(2) ≈ 2.872468
     With h =   0.05: y(2) ≈ 2.872468
     With h = 0.025:   y(2) ≈ 2.872468

22. With h =   0.2:   y(2) ≈ 7.326761
     With h =   0.1:   y(2) ≈ 7.328452
     With h =   0.05: y(2) ≈ 7.328971
     With h = 0.025:   y(2) ≈ 7.329134

23. With h =   0.2:   y(1) ≈ 1.230725
     With h =   0.1:   y(1) ≈ 1.230731
     With h =   0.05: y(1) ≈ 1.230731
     With h = 0.025:   y(1) ≈ 1.230731

24. With h =   0.2:   y(1) ≈ 1.000000
     With h =   0.1:   y(1) ≈ 1.000000
     With h =   0.05: y(1) ≈ 1.000000
     With h = 0.025:   y(1) ≈ 1.000000

25. With step size h = 0.2 we get the following results:

t	v	s
5	100.16	319.89
10	104.98	839.56
15	105.05	1364.72
20	105.05	1889.95

At the terminal velocity of 105.05 ft/sec it will require
about 77.2 seconds to fall the remaining 8110.05 ft to
the ground. Hence the total time of descent is about 1 min
37 sec.

26. With step size h = 0.1 we get the following results:

t	v	Approx s	Actual s
5	77.12	265.06	265.00
10	79.95	661.49	661.44
15	80.00	1061.43	1061.37
20	80.00	1461.43	1461.37

At the terminal velocity of 80 ft/sec it will require about 107 seconds to fall the remaining 8539 ft to the ground. Hence the total time of descent is about 2 min 7 sec.

27. (a) The car's terminal velocity is 100 ft/sec ≈ 68 mph
    (b) After 10 sec it has traveled about 368 ft and has velocity v(10) ≈ 63.21 ft/sec ≈ 43 mph.
    (c) It reaches a velocity of 88 ft/sec in about 21 seconds.

28. (a) The car's terminal velocity is 100 ft/sec ≈ 68 mph
    (b) After 10 sec it has traveled about 434 ft and has velocity v(10) ≈ 76.16 ft/sec ≈ 52 mph.
    (c) It reaches a velocity of 88 ft/sec in slightly under 14 seconds.

29. The ball reaches a maximum height of about 302.5 ft in about 4.05 sec. It requires about 4.7 sec to fall back to the ground, and hits with speed of about 120 ft/sec.

30. The ball reaches a maximum height of about 294 ft in about 4.08 sec. It requires about 4.5 sec to fall back to the ground, and hits with speed of about 119 ft/sec.

SECTION 6.4

SYSTEMS OF DIFFERENTIAL EQUATIONS

In Problems 1-8 we give the approximate and actual values x(0.2) and y(0.2) of the solution at t = 0.2.

1. Euler values: x(0.2) ≈ 0.8800,   y(0.2) ≈ 2.5000
   Improved Euler: x(0.2) ≈ 0.9600,   y(0.2) ≈ 2.6000
   Runge-Kutta: x(0.2) ≈ 1.0027,   y(0.2) ≈ 2.6401
   Actual values: x(0.2) ≈ 1.0034,   y(0.2) ≈ 2.6408

2. Euler values: x(0.2) ≈ 0.8100,   y(0.2) ≈ -0.8100
   Improved Euler: x(0.2) ≈ 0.8200,   y(0.2) ≈ -0.8200
   Runge-Kutta: x(0.2) ≈ 0.8187,   y(0.2) ≈ -0.8187
   Actual values: x(0.2) ≈ 0.8187,   y(0.2) ≈ -0.8187

3. Euler values: x(0.2) ≈ 2.8100,   y(0.2) ≈ 2.3100
   Improved Euler: x(0.2) ≈ 3.2200,   y(0.2) ≈ 2.6200
   Runge-Kutta: x(0.2) ≈ 3.6481,   y(0.2) ≈ 2.9407
   Actual values: x(0.2) ≈ 3.6775,   y(0.2) ≈ 2.9628

4. Euler values: x(0.2) ≈ 3.3100,   y(0.2) ≈ -1.6200
   Improved Euler: x(0.2) ≈ 3.8200,   y(0.2) ≈ -2.0400
   Runge-Kutta: x(0.2) ≈ 4.2274,   y(0.2) ≈ -2.4060
   Actual values: x(0.2) ≈ 4.2427,   y(0.2) ≈ -2.4205

5. Euler values: x(0.2) ≈ -0.5200,   y(0.2) ≈ 2.9200
   Improved Euler: x(0.2) ≈ -0.8400,   y(0.2) ≈ 2.4400
   Runge-Kutta: x(0.2) ≈ -0.5712,   y(0.2) ≈ 2.4485
   Actual values: x(0.2) ≈ -0.5793,   y(0.2) ≈ 2.4488

6. Euler values: x(0.2) ≈ -1.7600,   y(0.2) ≈ 4.6800
   Improved Euler: x(0.2) ≈ -1.9200,   y(0.2) ≈ 4.5600
   Runge-Kutta: x(0.2) ≈ -1.9029,   y(0.2) ≈ 4.4995
   Actual values: x(0.2) ≈ -1.9025,   y(0.2) ≈ 4.4999

7. Euler values: x(0.2) ≈ 3.1200,   y(0.2) ≈ 1.6800
   Improved Euler: x(0.2) ≈ 3.2400,   y(0.2) ≈ 1.7600
   Runge-Kutta: x(0.2) ≈ 3.2816,   y(0.2) ≈ 1.7899
   Actual values: x(0.2) ≈ 3.2820,   y(0.2) ≈ 1.7902

8. Euler values: x(0.2) ≈ 2.1600,   y(0.2) ≈ -0.6300
   Improved Euler: x(0.2) ≈ 2.5200,   y(0.2) ≈ -0.4600
   Runge-Kutta: x(0.2) ≈ 2.5320,   y(0.2) ≈ -0.3867
   Actual values: x(0.2) ≈ 2.5270,   y(0.2) ≈ -0.3889

In Problems 9-11 we give the Runge-Kutta approximate values with step sizes h = 0.1 and h = 0.05, and also the actual values.

9.   With h = 0.1:   x(1) ≈ 3.99261,   y(1) ≈ 6.21770
     With h = 0.05:   x(1) ≈ 3.99234,   y(1) ≈ 6.21768
     Actual values: x(1) ≈ 3.99232,   y(1) ≈ 6.21768

10. With h = 0.1:   x(1) ≈ 1.31498,   y(1) ≈ 1.02537
     With h = 0.05:   x(1) ≈ 1.31501,   y(1) ≈ 1.02538
     Actual values: x(1) ≈ 1.31501,   y(1) ≈ 1.02538

11. With h = 0.1:   x(1) ≈ -0.05832,   y(1) ≈ 0.56664
     With h = 0.05:   x(1) ≈ -0.05832,   y(1) ≈ 0.56665
     Actual values: x(1) ≈ -0.05832,   y(1) ≈ 0.56665

12. We first convert the given initial value problem to the two-dimensional problem

$$x' = y \qquad\qquad x(0) = y(0) = 0$$
$$y' = -x + \sin t.$$

Then with both step sizes h = 0.1 and h = 0.05 we get the actual value x(1) ≈ 0.15058 to 5 decimal places.

13. With y = x' we want to solve numerically the initial value problem

$$x' = y, \qquad\qquad x(0) = 0$$
$$y' = -32 - 0.04y, \qquad y(0) = 288.$$

When we run Program RK2DIM with step size h = 0.1 we find that the change of sign in the velocity v occurs as follows:

t	x	v
7.6	1050.2	+2.8
7.7	1050.3	-0.4

Thus the bolt attains a maximum height of about 1050 ft in about 7.7 sec.

14. Now we want to solve numerically the initial value problem

$$x' = y, \qquad\qquad x(0) = 0$$
$$y' = -32 - 0.0002y^2, \qquad y(0) = 288.$$

Running Program RK2DIM with step size h = 0.1, we find that the bolt attains a maximum height of about 1044 ft in about 7.8 sec. Note that these values are comparable to those found in Problem 13.

15. With y = x', and with x in miles and t in seconds, we want to solve numerically the initial value problem

$$x' = y$$
$$y' = -95485.5/(x^2 + 7920x + 15681600)$$
$$x(0) = 0, \qquad y(0) = 1.$$

We find (running RK2DIM with h = 1) that the projectile reaches a maximum height of about 83.83 miles in about 168 sec = 2 min 48 sec.

16. We need only run Program BASEBALL with the indicated initial inclination angles. Using step size h = 0.1 we get the following results:

Angle	Time	Range
40	5.0	352.9
45	5.4	347.2
50	5.8	334.2

We have listed the time to the nearest tenth of a second, but have interpolated to find the range in feet.

17. The data in Problem 16 indicate that the range increases when the initial angle is decreased below 45°. The further data

Angle	Range
41.0	352.1
40.5	352.6
40.0	352.9
39.5	352.8
39.0	352.7
35.0	350.8

indicate that a maximum range of about 353 ft is attained with $\alpha \approx 40^\circ$.

18. We "shoot" for the proper inclination angle by running Program BASEBALL (with h = 0.1) as follows:

Angle	Range
60	.287.1
58	298.5
57.5	301.1

Thus we get a range of 300 ft with an initial angle just under 57.5°.

19. First we run Program BASEBALL (with h= 0.1) with $v_0 = 250$ ft/sec and obtain the following results:

t	x	y
5.0	457.43	103.90
6.0	503.73	36.36

Interpolation gives x = 494.4 when y = 50. Then a run with $v_0 = 255$ ft/sec gives the following results:

t	x	y
5.5	486.75	77.46
6.0	508.86	41.62

Finally a run with $v_0 = 253$ ft/sec gives these results:

t	x	y
5.5	484.77	75.44
6.0	506.82	39.53

Now x ≈ 500 ft when y = 50 ft. Thus Babe Ruth's home run
ball had an initial velocity of 253 ft/sec.

20. A run of Program BASEBALL with h = 0.1 and with the given
data yields the following results:

t	x	y	v	α
5.5	989	539	162	+0.95
5.6	1005	539	161	-0.18
.	.	.	.	.
.	.	.	.	.
.	.	.	.	.
.	.	.	.	.
11.5	1868	16	214	-52
11.6	1881	-1	216	-53

The first two lines of data above indicate that the crossbow
bolt attains a maximum height of about 1005 ft in about
5.6 sec. About 6 sec later (total time 11.6 sec) it hits
the ground, having traveled about 1880 ft horizontally.

21. A run with h = 0.1 indicates that the projectile has a
range of about 21,400 ft ≈ 4.05 mi and a flight time of
about 46 sec. It attains a maximum height of about 8970
ft in about 17.5 sec. At time t ≈ 23 sec it has its
minimum velocity of about 368 ft/sec. It hits the ground
(t ≈ 46 sec) at an angle of about $77°$ with a velocity of
about 518 ft/sec.

22. The required Program KEPLER is listed below. Note that only
in input and output, and in the definitions of the functions
f and g, does it differ materially from Program BASEBALL.

```
100 REM--Program KEPLER
110 REM
120 REM--Initialization:
130 REM
140      DEFINT K-N : DEFDBL A-H,P-Z
160      DEF FNF(X,Y) = - X/(X*X + Y*Y)^(3/2)
170      DEF FNG(X,Y) = - Y/(X*X + Y*Y)^(3/2)
180      INPUT "Initial t"; T
190      INPUT "Initial x,y"; X,Y
200      INPUT "Initial p,q"; P,Q
220      INPUT "Step size h"; H
230      INPUT "Number of steps"; M
```

```
240        INPUT "Print step k"; K
250        PRINT T,X,Y
280 REM
290 REM--Runge-Kutta iteration:
300 REM
310        FOR N = 1 TO M
320            F1 = FNF(X,Y)
330            G1 = FNG(X,Y)
340            F2 = FNF(X + H*F1/2,Y + H*G1/2)
350            G2 = FNG(X + H*F1/2,Y + H*G1/2)
360            F3 = FNF(X + H*F2/2,Y + H*G2/2)
370            G3 = FNG(X + H*F2/2,Y + H*G2/2)
380            F4 = FNF(X + H*F3,  Y + H*G3)
390            G4 = FNG(X + H*F3,  Y + H*G3)
400            DP = (F1 + 2*F2 + 2*F3 + F4)/6
410            DQ = (G1 + 2*G2 + 2*G3 + G4)/6
420            X  = X + P*H + .5*DP*H*H
430            Y  = Y + Q*H + .5*DQ*H*H
440            P  = P + DP*H
450            Q  = Q + DQ*H
460            T = T + H
470            IF INT(N/K) = N/K THEN PRINT T,X,Y
480        NEXT N
490 REM
500        END
```

CHAPTER 7

NONLINEAR DIFFERENTIAL EQUATIONS AND SYSTEMS

SECTION 7.1

INTRODUCTION TO STABILITY

1.   Explicit solution: $x(t) = x_0 e^{-t}$

Critical point: $x = 0$

For all x_0, $x(t) \to 0$ as $t \to \infty$, so $x = 0$ is a *stable* critical point.

2.   Explicit solution: $x(t) = 2 + (x_0 - 2)e^t$

Critical point: $x = 2$

If $x_0 \neq 2$ then $|x(t)| \to \infty$ as $t \to \infty$. Hence $x = 2$ is an *unstable* critical point.

3.   Explicit solution: $x(t) = x_0 / [x_0 + (1 - x_0)e^{-t}]$

Critical points: $x = 0$ and $x = 1$

If $x_0 > 0$ then $x(t) \to 1$ as $t \to \infty$, but if $x_0 < 0$ then $x(t) \to -\infty$ as $t \to \ln[x_0/(x_0 - 1)]$. Hence $x = 1$ is a *stable* critical point, but $x = 0$ is an *unstable* critical point.

4.   Explicit solution: $x(t) = 2x_0 / [x_0 + (2 - x_0)e^{2t}]$

Critical points: x = 0   and   x = 2

If x_0 < 0   or   0 < x_0 < 2   then   x(t) —> 0   as   t —> ∞.
Hence   x = 0   is a *stable* critical point.   But if x_0 > 2
then   x(t) —> ∞   as   t —> (1/2) ln[x_0/(x_0 - 2)].   Hence   x =
2   is an *unstable* critical point.

5.   Explicit solution: x(t) = x_0/[(x_0 + 1)e^{-t} - x_0]

Critical points: x = 0   and   x = -1

If x_0 > 0   then   x(t) —> +∞   as   t —> ln(x_0 + 1)/x_0,   but
if x_0 < 0   then   x(t) —> -1   as   t —> ∞.   Hence   x = 0
is an *unstable* critical point, whereas   x = -1   is a *stable*
critical point.

6.   Explicit solution: x(t) = [x_0 - t(x_0 - 1)]/[1 - t(x_0 - 1)]

Critical point: x = 1

If x_0 < 1   then   x(t) —> 1   as   t —> ∞,   but if x_0 > 1
then   x(t) —> ∞   as   t —> 1/(x_0 - 1).   Hence   x = 1   is an
unstable critical point.

7.   Explicit solution:

$$x(t) = [1 + x_0 - (1 - x_0)e^{-2t}]/[1 + x_0 + (1 - x_0)e^{-2t}]$$

Critical points: x = +1   and   x = -1

If x_0 > -1   then   x(t) —> +1   as   t —> ∞, but if x_0 < -1
then   x(t) —> -∞   as   t —> (1/2) ln[(x_0 - 1)/(x_0 + 1)].
Hence   x = +1   is a *stable* critical point, but   x = -1   is
an *unstable* critical point.

8. Explicit solution:

$$x(t) = 2[(x_0 + 2) + (x_0 - 2)e^{4t}]/[(x_0 + 2) - (x_0 - 2)e^{4t}]$$

Critical points: $x = -2$ and $x = 2$
If $x_0 < 2$ then $x(t) \rightarrow -2$ as $t \rightarrow \infty$, but if $x_0 > 2$
then $x(t) \rightarrow \infty$ as $t \rightarrow (1/4)\ln[(x_0 + 2)/(x_0 - 2)]$. Hence
$x = -2$ is a *stable* critical point, but $x = 2$ is an
unstable critical point.

9. Explicit solution:

$$x(t) = [x_0 + 2 + 2(x_0 - 1)e^{-3t}]/[x_0 + 2 - (x_0 - 1)e^{-3t}]$$

Critical points: $x = 1$ and $x = -2$

If $x_0 > -2$ then $x(t) \rightarrow 1$ as $t \rightarrow \infty$, but if $x_0 < -2$
then $x(t) \rightarrow -\infty$ as $t \rightarrow (1/3)\ln[(x_0 - 1)/(x_0 + 2)]$.
Hence $x = 1$ is a *stable* critical point, but $x = -2$ is an
unstable critical point.

10. Implicit solution: $x^2 = (x_0)^2/[(x_0)^2 + (1 - (x_0)^2)e^{-2t}]$

Critical points: $x = -1, x = 0, x = +1$

If $x_0 < 0$ then $x(t) \rightarrow -1$ as $t \rightarrow \infty$, but if $x_0 > 0$
then $x(t) \rightarrow +1$ as $t \rightarrow \infty$. Hence $x = 0$ is an *unstable*
critical point, but $x = -1$ and $x = +1$ are *stable*
critical points.

11. Explicit solution:

$$x(t) = [2(x_0 - 3) - 3(x_0 - 2)e^{-t}]/[(x_0 - 3) - (x_0 - 2)e^{-t}]$$

Critical Points: $x = 2$ and $x = 3$

If $x_0 < 3$ then $x(t) \rightarrow 2$ as $t \rightarrow \infty$, but if $x_0 > 3$
then $x(t) \rightarrow +\infty$ as $t \rightarrow \ln[(x_0 - 2)/(x_0 - 3)]$. Hence

$x = 2$ is a *stable* critical point, but $x = 3$ is an *unstable* critical point.

12. Explicit solution:

$$x(t) = 2 + (x_0 - 2)/[1 + 2t(2 - x_0)^2]^{1/2}$$

Critical point: $x = 2$

For all x_0 it is clear that $x(t) \longrightarrow 2$ as $t \longrightarrow \infty$, so $x = 2$ is a *stable* critical point.

13. (a) With $k = 0$ the solution of $x' = -x^3$ is

$$x(t) = x_0/(2x_0^2 t + 1)^{1/2}.$$

With $k = -a^2 < 0$ the solution of $x' = kx - x^3$ is given by

$$x^2 = [a^2 x_0^2 \exp(-2a^2 t)]/[a^2 + x_0^2 - x_0^2 \exp(-2a^2 t)].$$

In either case it is clear that $x(t) \longrightarrow 0$ as $t \longrightarrow \infty$, so if $k \le 0$ the single critical point $x = 0$ is *stable*.

(b) If $k = +a^2 > 0$ then the solution is given by

$$x(t) = ax_0/[x_0^2 + (a^2 - x_0^2)\exp(-2a^2 t)]^{1/2}.$$

As $t \longrightarrow \infty$, it follows that $x(t) \longrightarrow a$ if $x_0 > 0$, while $x(t) \longrightarrow -a$ if $x_0 < 0$. Consequently the critical points $x = \pm a$ are *stable* while $x = 0$ is *unstable*.

14. The solution of $x' = ax - bx^2$, $x(0) = x_0$ is

$$x(t) = ax_0/[bx_0 + (a - bx_0)e^{-at}].$$

If $x_0 > 0$ then it follows that $x(t) \longrightarrow a/b$ as $t \longrightarrow \infty$. But if $x_0 < 0$ then it follows that $x(t) \longrightarrow -\infty$ as $t \longrightarrow$

$(1/a)\ln[(bx_0 - a)/bx_0]$. Hence the critical point $x = a/b$ is *stable*, while the critical point $x = 0$ is *unstable*.

15. The solution of $x' = bx^2 - ax$, $x(0) = x_0$ is

$$x(t) = ax_0e^{-at}/[(a - bx_0) + bx_0e^{-at}].$$

If $x_0 < a/b$ then $a - bx_0 > 0$, so it follows that $x(t) \longrightarrow 0$ as $t \longrightarrow \infty$. But if $x_0 > a/b$ then $x(t) \longrightarrow \infty$ as $t \longrightarrow (1/a)\ln[bx_0/(bx_0 - a)]$. Hence the critical point $x = a/b$ is *unstable*, while the critical point $x = 0$ is *stable*.

16. Now the differential equation is

$$x' = (a - k)x - bx^2$$

(a) If $0 < k < a$ then $a - k > 0$, so by the computation in Problem 14 the new limiting population is $M = (a - k)/b$.

(b) If $k = a$ then

$$x(t) = 1/[bt + (1/x_0)].$$

If $k > a$ then

$$x(t) = Ae^{-ct}/(1 - rAe^{-ct})$$

where $c = k - a > 0$, $r = b/c$, and $A = x_0/(1 + rx_0)$. In either case it is clear that $x(t) \longrightarrow 0$ as $t \longrightarrow \infty$.

17. If $4bk = a^2$ then

$$x' = - b(x - a/2b)^2.$$

It follows by separation of variables that

$$x(t) = (a/2b) + 1/(bt + C),$$

so $x(t) \to a/2b$ as $t \to \infty$.

18.   If $4bk > a^2$ let $\gamma^2 = (4kb - a^2)/4b^2 > 0$ and substitute $u = x - a/2b$.   Then the equation

$$u' = -b(u^2 + \gamma^2)$$

has solution

$$u = \gamma \tan(C - b\gamma t).$$

Then $x = 0$ when $\gamma \tan(C - b\gamma t) = -a/2b$.

19.   If $4bk < a^2$ let $\gamma^2 = (a^2 - 4bk)/4b^2 > 0$ and substitute $u = x - a/2b$.   Then the equation

$$u' = -b(u^2 - \gamma^2)$$

has solution $u(t)$ satisfying

$$u - \gamma = C(u + \gamma)e^{-2b\gamma t},$$

so $u = x - a/2b \to \gamma$ as $t \to \infty$.

20.   The solution of the initial value problem

$$x' = b(x - \alpha)(x - \beta), \quad x(0) = x_0$$

is

$$x(t) = \frac{\alpha(x_0 - \beta) + \beta(x_0 - \alpha)e^{rt}}{(x_0 - \beta) - (x_0 - \alpha)e^{rt}}$$

where $r = (a^2 + 4bk)^{1/2}$.   If $x_0 > \alpha$ then the quantities within parentheses are all positive,   so $x(t) \to \infty$ as $t \to (1/r)\ln[(x_0 - \beta)/(x_0 - \alpha)]$.   But if $x_0 < \alpha$ then $x(t) \to -\beta < 0$ as $t \to \infty$,   so $x(t) \to 0$ in a finite period of time.

STABILITY AND THE PHASE PLANE

1.   The only critical point is   (0,0).

2.   All points on the line   x = 2y

3.   (-2,-3)

4.   (0,0)

5.   Suppose   x(2 - y) = 0   and   y(x - 3) = 0.   Then   x = 0
      implies   y = 0,   and conversely.   If   x   and   y   are both
      nonzero, then cancellation gives   2 - y = x - 3 = 0.   Hence
      there are two critical points,   (0,0)   and   (3,2).

6.   (0,0)   and   (1,2)

7.   There are infinitely many critical points -- every point of
      the form   (nπ,0)   with   n   an integer.

8.   If   x = 0   then the first equation yields   y = 0.   Similarly
      y = 0   in the second equation gives   x = 0.   If   x   and   y
      are both nonzero then we can solve for

$$x^2 + y^2 = 1 + (y/x) = 1 - (x/y).$$

      But   y/x = - x/y   implies that $x^2 + y^2 = 0$,   so   x = y = 0,
      after all.   Hence   (0,0)   is the only critical point.

In each of Problems 9-12 we need only set   x' = x" = 0   and solve
the resulting equation for   x.

9.   x(t) = 0,   x(t) = 2,   x(t) = -2

10.   x(t) = 0

11. There are infinitely many equilibrium solutions: $x(t) = n\pi$ for any integer n.

12. $x(0) = 0$

13. Solution: $x(t) = x_0 e^{-2t}$, $y(t) = y_0 e^{-2t}$

    The origin is a stable proper node like the one shown in Figure 7.4.

14. Solution: $x(t) = x_0 e^{2t}$, $y(t) = y_0 e^{-2t}$

    The origin is an unstable saddle point like the one illustrated in Figure 7.6, except with all the arrows reversed.

15. Solution: $x(t) = x_0 e^{-2t}$, $y(t) = y_0 e^{-t}$

    The origin is a stable improper node like the one shown in Figure 7.5, except that the trajectories consist of the x-axis and parabolas of the form $x = ky^2$.

16. Solution: $x(t) = x_0 e^{t}$, $y(t) = y_0 e^{3t}$

    The origin is an unstable improper node. The trajectories consist of the y-axis and curves of the form $y = kx^3$, departing from the origin.

17. Solution: $x(t) = A \cos t + B \sin t$
              $y(t) = B \cos t - A \sin t$

    The origin is a stable center. The trajectories are clockwise-oriented circles centered at the origin.

18. Solution: $x(t) = A \cos 2t + B \sin 2t$,
              $y(t) = -2B \cos 2t + 2A \sin 2t$

    The origin is a stable center like the one illustrated in

Figure 7.7, except that the arrows point in a
counter-clockwise direction, and the vertical semiaxis of
each ellipse is twice its horizontal semiaxis.

19. Solution: $x(t) = A \cos 2t + B \sin 2t$,
$$y(t) = B \cos 2t - A \sin 2t$$

The origin is a stable center, and the trajectories are
clockwise-oriented circles centered at $(0,0)$.

20. Solution: $x(t) = e^{-2t}(A \cos t + D \sin t)$,
$$y(t) = e^{-2t}[(-2A + B)\cos t - (A + 2B)\sin t]$$

The origin is an asymptotically stable spiral point with
trajectories approaching $(0,0)$ as in Figure 7.8.

23. Origin $(0,0)$ and the circles $x^2 + y^2 = C > 0$.

24. Origin $(0,0)$ and the hyperbolas $y^2 - x^2 = C$.

25. Origin $(0,0)$ and the ellipses $x^2 + 4y^2 = C > 0$.

26. Origin $(0,0)$ and the ovals of the form $x^4 + y^4 = C > 0$.

27. If $\varphi(t) = x(t + \gamma)$ and $\theta(t) = y(t + \gamma)$ then

$$\varphi'(t) = x'(t + \gamma) = y(t + \gamma) = \theta(t),$$

but
$$\theta(t) = y'(t + \gamma)$$
$$= x(t + \gamma) \cdot (t + \gamma)$$
$$= t\varphi(t) + \gamma\varphi(t) \neq t\varphi(t).$$

28. If $\varphi(t) = x(t + \gamma)$ and $\theta(t) = y(t + \gamma)$ then

$$\varphi'(t) = x'(t + \gamma)$$
$$= F(x(t + \gamma),y(t + \gamma))$$
$$= F(\varphi(t),\theta(t)),$$

and

$$\theta'(t) = G(\varphi(t), \theta(t))$$

similarly. Therefore $\varphi(t)$ and $\theta(t)$ satisfy the given differential equations.

SECTION 7.3

LINEAR AND ALMOST LINEAR SYSTEMS

In Problems 1-10 we first find the roots λ_1 and λ_2 of the characteristic equation of the given system. We can then read the type and stability of the critical point $(0,0)$ from Theorem 1 and the table in Figure 7.17.

1. $\lambda_1 = -1$, $\lambda_2 = -3$ so $(0,0)$ is an asymptotically stable node.

2. $\lambda_1 = 2$, $\lambda_2 = 3$ so $(0,0)$ is an unstable improper node.

3. $\lambda_1 = -1$, $\lambda_2 = 3$ so $(0,0)$ is an unstable saddle point.

4. $\lambda_1 = -2$, $\lambda_2 = 4$ so $(0,0)$ is an unstable saddle point.

5. $\lambda_1 = \lambda_2 = -1$ so $(0,0)$ is an asymptotically stable node.

6. $\lambda_1 = \lambda_2 = 2$ so $(0,0)$ is an unstable node.

7. $\lambda_1, \lambda_2 = 1 \pm 2i$ so $(0,0)$ is an unstable spiral point.

8. $\lambda_1, \lambda_2 = -2 \pm 3i$ so $(0,0)$ is an asymptotically stable spiral point.

9. $\lambda_1, \lambda_2 = \pm 2i$ so $(0,0)$ is a stable (but not asymptotically stable) center.

10. λ_1, $\lambda_2 = \pm 3i$ so $(0,0)$ is a stable (but not asymptotically stable) center.

11. The substitution $u = x - 2$, $v = y - 1$ transforms the given system to the system

$$u' = u - 2v, \qquad v' = 3u - 4v$$

with characteristic roots $\lambda_1 = -1$, $\lambda_2 = -2$. Hence $(2,-1)$ is an asymptotically stable node.

12. The substitution $u = x - 2$, $v = y + 3$ transforms the given system to the system

$$u' = u - 2v, \qquad v' = u + 4v$$

with characteristic roots $\lambda_1 = 2$, $\lambda_2 = 3$. Hence $(2,-3)$ is an unstable improper node.

13. The substitution $u = x - 2$, $v = y - 2$ transforms the given system to the system

$$u' = 2u - v, \qquad v' = 3u - 2v$$

with characteristic roots $\lambda_1 = 1$, $\lambda_2 = -1$. Hence $(2,2)$ is an unstable saddle point.

14. The substitution $u = x - 3$, $v = y - 4$ transforms the given system to the system

$$u' = u + v, \qquad v' = 3u - v$$

with characteristic roots $\lambda_1 = -2$, $\lambda_2 = 2$. Hence $(3,4)$ is an unstable saddle point.

15. The substitution $u = x - 1$, $v = y - 1$ transforms the given system to the system

$$u' = u - v, \qquad\qquad v' = 5u - 3v$$

with characteristic roots λ_1, $\lambda_2 = -1 \pm i$. Hence $(1,1)$ is an asymptotically stable spiral point.

16. The substitution $u = x - 3$, $v = y - 2$ transforms the given system to the system

$$u' = u - 2v, \qquad\qquad v' = u + 3v$$

with characteristic roots λ_1, $\lambda_2 = 2 \pm i$. Hence $(3,2)$ is an unstable spiral point.

17. The substitution $u = x - 5/2$, $v = y + 1/2$ transforms the given system to the system

$$u' = u - 5v, \qquad\qquad v' = u - v$$

with characteristic roots λ_1, $\lambda_2 = \pm\, 2i$. Hence $(5/2, -1/2)$ is a stable center.

18. The substitution $u = x + 2$, $v = y + 1$ transforms the given system to the system

$$u' = 4u - 5v, \qquad\qquad v' = 5u - 4v$$

with characteristic roots λ_1, $\lambda_2 = \pm\, 3i$. Hence $(-2, -1)$ is a stable (but not asymptotically stable) center.

In Problems 19-28 we first find the characteristic roots λ_1 and λ_2 of the associated linear system, and then apply Theorem 2 to determine as much as we can about the type and stability of the critical point $(0,0)$ of the given almost linear system.

19. $\lambda_1 = -2$, $\lambda_2 = -3$ so $(0,0)$ is an asymptotically stable node.

20. $\lambda_1 = 1$, $\lambda_2 = 4$ so $(0,0)$ is an unstable improper node.

21. $\lambda_1 = -3$, $\lambda_2 = 2$ so (0,0) is an unstable saddle point.

22. λ_1, $\lambda_2 = \pm 3i$ so (0,0) is either a center or a spiral point, but its stability is not determined by Theorem 2.

23. λ_1, $\lambda_2 = -2 \pm 2i$ so (0,0) is an asymptotically stable spiral point.

24. λ_1, $\lambda_2 = 1 \pm 3i$ so (0,0) is an unstable spiral point.

25. $\lambda_1 = \lambda_2 = -1$ so (0,0) is an asymptotically stable node or spiral point.

26. $\lambda_1 = \lambda_2 = 1$ so (0,0) is an unstable critical point that is either a node or a spiral point.

27. λ_1, $\lambda_2 = \pm i$ so (0,0) is either a center or a spiral point, but its stability is not determined by Theorem 2.

28. λ_1, $\lambda_2 = \pm 2i$ so (0,0) is either a center or a spiral point, but its stability is not determined by Theorem 2.

29. The critical points of the given system are (0,0) and (1,1). At (0,0) the characteristic roots are $\lambda_1 = -1$, $\lambda_2 = 1$ so (0,0) is an unstable saddle point. The substitution $u = x - 1$, $v = y - 1$ transforms the given system to the almost linear system

$$u' = u - v, \qquad v' = 2u + v + u^2$$

whose linearization has characteristic roots λ_1, $\lambda_2 = \pm i$. Hence (1,1) is either a center or a spiral point, but its stability is indeterminate.

30. The critical points of the given system are (1,1) and (-1,1). The substitution $u = x - 1$, $v = y - 1$ transforms it to the almost linear system

$$u' = v, \qquad v' = 2u - v + u^2$$

whose linearized system has characteristic roots $\lambda_1 = -2$, $\lambda_2 = 1$. Hence $(1,1)$ is an unstable saddle point of the given system. The substitution $u = x + 1$, $v = y - 1$ transforms it to the almost linear system

$$u' = v, \qquad v' = -2u + v + u^2$$

whose linearized system has characteristic roots λ_1, $\lambda_2 = (-1 \pm i\sqrt{7})/2$. Hence $(-1,1)$ is an asymptotically stable spiral point of the given system.

31. The critical points of the given system are $(1,1)$ and $(-1,-1)$. The substitution $u = x - 1$, $v = y - 1$ transforms it to the almost linear system

$$u' = 2v + v^2, \qquad v' = 3u - v + 3u^2 + u^3$$

whose linearization has characteristic roots $\lambda_1 = -3$, $\lambda_2 = 2$. Hence $(1,1)$ is an unstable saddle point of the given system. The substitution $u = x + 1$, $v = y + 1$ transforms it to the almost linear system

$$u' = -2v + v^2, \quad v' = 3u - v - 3u^2 + u^3$$

whose linearization has characteristic roots λ_1, $\lambda_2 = (-1 \pm i\sqrt{23})/2$. Hence $(-1,-1)$ is an asymptotically stable spiral point.

32. The critical points of the given system are $(2,1)$ and $(-2,-1)$. The substitution $u = x - 2$, $v = y - 1$ transforms it to the almost linear system

$$u' = u + 2v + uv, \qquad v' = u - 2v$$

whose linearized system has characteristic roots λ_1, $\lambda_2 = (-1 \pm \sqrt{17})/2$ of opposite sign. Hence $(2,1)$ is an

unstable saddle point of the given system. The substitution
$u = x + 2$, $v = y + 1$ transforms it to the almost linear
system

$$u' = - u - 2v + uv, \qquad v' = u - 2v$$

whose linearized system has characteristic roots λ_1, $\lambda_2 = (-3 \pm i \sqrt{7})/2$. Hence $(-2,-1)$ is an asymptotically stable spiral point of the given system.

34. The characteristic equation of the given linear system is

$$(\lambda + 1)^2 - h = 0.$$

(a) If $h = 0$ then the characteristic roots $\lambda_1 = \lambda_2 = -1$ are equal and negative, so $(0,0)$ is an asymptotically stable node.

(b) If $h < 0$ then λ_1, $\lambda_2 = -1 \pm i \sqrt{(-h)}$, so $(0,0)$ is an asymptotically stable spiral point.

(c) If $0 < h < 1$ then λ_1, $\lambda_2 = -1 \pm \sqrt{h}$ are both negative, so $(0,0)$ is an asymptotically stable improper node.

36. The substitution $y = vx$ in the homogeneous first order equation

$$\frac{dy}{dx} = \frac{y(2x^3 - y^3)}{x(x^3 - 2y^3)}$$

yields

$$x \frac{dv}{dx} = - \frac{v^4 + v}{2v^3 - 1}.$$

Separating the variables and integrating by partial fractions, we get

$$\int \left[-\frac{1}{v} + \frac{1}{v+1} + \frac{2v-1}{v^2-v+1} \right] dv = -\int \frac{dx}{x}$$

$$\ln\,[(\,v+1)(v^2-v+1)] = \ln v - \ln x + \ln C$$

$$(v+1)(v^2-v+1) = Cv/x$$

$$v^3 + 1 = Cv/x.$$

Finally the replacement $v = y/x$ yields

$$x^3 + y^3 = Cxy.$$

SECTION 7.4

ECOLOGICAL APPLICATIONS - PREDATORS AND COMPETITORS

1. The effect of using the insecticide is to replace c by $c + f$ and a by $a - f$ in Equations (1) while leaving b and d unchanged. Hence the new harmful population is

$$(c+f)/d > c/d = x_E,$$

and the new benign population is

$$(a-f)/b < a/b = y_E.$$

Problems 2-6 deal with the two competing populations that are discussed in Example 2 in the text. Problem 2 records the fact that the critical points of the system

$$x' = 60x - 3x^2 - 4xy$$
$$y' = 42x - 3y^2 - 2xy$$

are $(0,0)$, $(0,14)$, $(20,0)$, and $(12,6)$, as found in Example 1 of Section 7.2.

3.   At (0,0) the linearized system is

$$x' = 60x \qquad\qquad y' = 42y$$

with general solution

$$x(t) = Ae^{60t}, \qquad y(t) = Be^{42t}.$$

Thus (0,0) is an unstable node with linearized trajectories of the form $y = kx^{7/10}$ tangent to the y-axis.

4.   At (0,14) the linearized system

$$u' = 4u, \qquad\qquad v' = -28u - 42v$$

has eigenvalues $\lambda_1 = 4$ and $\lambda_2 = -42$, so (0,14) is an unstable saddle point of the original almost linear system.

5.   At (20,0) the linearized system

$$u' = -60u - 80v, \qquad v' = 2v$$

has eigenvalues $\lambda_1 = 2$ and $\lambda_2 = -60$, so (20,0) also is an unstable saddle point.

6.   At (12,6) the linearized system

$$u' = -36u - 48v, \qquad v' = -12u - 18v$$

has eigenvalues $\lambda_1, \lambda_2 = -27 \pm 3\sqrt{73}$ that are both negative, so (12,6) is a stable node.

Problems 7-11 deal with the system

$$x' = 5x - x^2 - xy,$$
$$y' = xy - 2y.$$

Problem 7 records the fact that the critical points of this system are (0,0), (5,0), and (2,3).

8.   At (0,0) the linearized system is

$$x' = 5x, \qquad\qquad y' = -2y$$

with general solution $x(t) = Ae^{5t}$, $y(t) = Be^{-2t}$. Thus (0,0) is an unstable saddle point of both the linearized system and the original system.

9.   To analyze the critical point (5,0) we substitute $u = x - 5$, $v = y$ and get the transformed system

$$u' = -5u - 5v - u^2 - uv,$$
$$v' = 3v + uv.$$

The associated linear system has eigenvalues $\lambda_1 = -5$, $\lambda_2 = 3$ so (5,0) is an unstable saddle point of the original system.   The general solution of the linearized system is

$$u = (-5A/8)e^{3t} + Be^{-5t}, \qquad v = Ae^{3t}.$$

With A = 0 we get trajectories entering along the x-axis, and with B = 0 we get trajectories departing along the line through (5,0) with slope -8/5.

10.   To analyze the critical point (2,3) we substitute $u = x - 2$, $v = y - 3$ and get the transformed system

$$u' = -2u - 2v - u^2 - uv,$$
$$v' = 3u - uv.$$

The associated linear system has eigenvalues λ_1, $\lambda_2 = -1 \pm i\sqrt{5}$, so (2,3) is an asymptotically stable spiral point.

Problems 12-17 deal with the system

$$x' = x^2 - 2x - xy,$$
$$y' = y^2 - 4y + xy.$$

Problem 12 records the fact that the critical points of this system are $(0,0)$, $(0,4)$, $(2,0)$, and $(3,1)$.

13. At $(0,0)$ the linearized system is

$$x' = -2x, \qquad y' = -4y$$

with general solution $x(t) = Ae^{-2t}$, $y(t) = Be^{-4t} = kx^2$. Hence the critical point $(0,0)$ is a stable improper node.

14. To analyze the critical point $(0,4)$ we substitute $u = x$, $v = y - 4$ and get the transformed system

$$u' = -6u + u^2 - uv,$$
$$v' = 4u + 4v + uv - v^2.$$

The associated linear system has eigenvalues $\lambda_1 = -6$, $\lambda_2 = 4$ so $(0,4)$ is an unstable saddle point of the original system. The general solution of the linearized system is

$$u = Ae^{-6t}, \qquad v = (-2A/5)e^{-6t} + Be^{4t}.$$

With $A = 0$ we get trajectories departing along the y-axis, and with $B = 0$ we get trajectories entering along the line through $(0,4)$ with slope $-2/5$.

15. To analyze the critical point $(2,0)$ we substitute $u = x - 2$, $v = y$ and get the transformed system

$$u' = 2u - 2v + u^2 - uv,$$
$$v' = -2v + uv + v^2.$$

The associated linear system has eigenvalues $\lambda_1 = -2$, $\lambda_2 = 2$ so $(2,0)$ is an unstable saddle point of the original system. The general solution of the linearized system is

$$u = Ae^{2t} + (B/2)\,e^{-2t}, \qquad\qquad v = Be^{-2t}.$$

With B = 0 we get trajectories departing along the x-axis, and with A = 0 we get trajectories entering along the line through (2,0) with slope 2.

16. To analyze the critical point (3,1) we substitute u = x − 3, v = y − 1 and get the transformed system

$$u' = 3u - 3v + u^2 - uv,$$
$$v' = u + v + uv + v^2.$$

The associated linear system has eigenvalues λ_1, λ_2 = 2 ± i√2, so the point (3,1) is an unstable spiral point.

20. We want to solve the system

$$B' = 0.00004\ BR - \qquad 0.02\ B$$
$$R' = \qquad 0.05\ R\ - 0.0004\ BR$$

with initial conditions B(0) = 100, R(0) = 600.
We assume that time is measured in days. The Runge-Kutta method yields the following approximate data:

t	B	R
0	100.00	600.00
50	145.03	668.37
100	147.63	373.14
150	107.65	359.40
200	99.82	597.32
250	144.59	670.74
300	147.96	374.66
350	107.92	358.30
400	99.65	594.63

It appears that the values of $B(t)$ and $R(t)$ return to approximately their initial values $B_0 = 100$ and $R_0 = 600$ after about 200 days and again after about 400 days. Near $t = 200$ and $t = 400$ we have the following approximate values:

t	B	R
195	98.20	567.00
200	99.82	597.32
205	102.09	626.81
395	98.08	564.29
400	99.65	594.63
405	101.86	624.24

By averaging the four available linear interpolations we estimate that the period is $T \approx 200.43$ days.

SECTION 7.5

NONLINEAR MECHANICAL SYSTEMS

In each of Problems 1-4 we need only substitute the familiar power series for the exponential, sine, and cosine functions, and then discard all higher-order terms. For each problem we give the corresponding linear system, the eigenvalues λ_1 and λ_2, and the type of this critical point.

1. $x' = -x + 2y$
   $y' = -x - 4y$
   $\lambda_1 = -2, \quad \lambda_2 = -3$
   Stable node

2. $x' = 2x + y$
   $y' = x + 2y$

   $\lambda_1 = 1, \quad \lambda_2 = 3$
   Unstable node

3. $x' = x + 2y$
    $y' = 8x + y$
    $\lambda_1 = -3, \quad \lambda_2 = 5$
    Unstable saddle node

4. $x' = x - 2y$
    $y' = 4x - 3y$

    because

    $\sin x \cos y = (x - x^3/3! + \cdots)(1 - y^2/2! + \cdots) = x + \cdots$.

    $\lambda_1, \lambda_2 = -1 \pm 2i$
    Stable spiral point

5. The critical points are of the form $(0, n\pi)$ where n is an integer. The linearized system at $(0, n\pi)$ is

    $$u' = -u \pm v$$
    $$v' = 2u$$

    where we take the plus sign if n is even, the minus sign if n is odd. If n is even the eigenvalues are $\lambda_1 = 1$ and $\lambda_2 = -2$, so $(0, n\pi)$ is an unstable saddle point. If n is odd the eigenvalues are $\lambda_1, \lambda_2 = (-1 \pm i\sqrt{7})/2$, so $(0, n\pi)$ is a stable spiral point.

6. The critical points are of the form $(n, 0)$ where n is an integer. The linearized system at $(n, 0)$ is

    $$u' = v$$
    $$v' = \pm \pi u - v$$

    where we take the plus sign if n is even, the minus sign if n is odd. The characteristic equation

    $$\lambda^2 + \lambda - \pi = 0$$

has one positive and one negative root, so (n,0) is an unstable saddle point if n is even. The equation

$$\lambda^2 + \lambda + \pi = 0$$

has complex conjugate roots with negative real part, so (n,0) is a stable spiral point if n is odd.

7. The critical points are of the form $(n\pi, n\pi)$ where n is an integer. The linearized system at $(n\pi, n\pi)$ is

$$u' = - u + v$$
$$v' = \pm 2u$$

where we take the plus sign if n is even, the minus sign if n is odd. If n is even the eigenvalues are $\lambda_1 = 1$ and $\lambda_2 = -2$, so $(n\pi, n\pi)$ is an unstable saddle point. If n is odd the eigenvalues are λ_1, $\lambda_2 = (-v \pm i\sqrt{7})/2$, so $(n\pi, n\pi)$ is a stable spiral point.

8. The critical points are of the form $(n\pi, 0)$ where n is an integer. The linearized system at $(n\pi, 0)$ is

$$u' = \pm 3u + v$$
$$v' = \pm u + 2v$$

where we take the plus signs if n is even, the minus signs if n is odd. If n is even then the eigenvalues λ_1, $\lambda_2 = (5 \pm \sqrt{5})/2$ are both positive, so $(n\pi, 0)$ is an unstable node. If n is odd then the eigenvalues λ_1, $\lambda_2 = (-1 \pm \sqrt{21})/2$ have different signs, so $(n\pi, 0)$ is an unstable saddle point.

9. The linearized system at the critical point $(n\pi, 0)$ is

$$u' = v$$
$$v' = \pm \omega^2 u - cv$$

where we take the plus sign if n is odd, the minus sign if n is even. If n is odd then the eigenvalues

$$\lambda_1, \lambda_2 = [- c \pm (c^2 + 4\omega^2)^{1/2}]/2$$

have different signs, so $(n\pi, 0)$ is an unstable saddle point. If n is even then the eigenvalues are

$$\lambda_1, \lambda_2 = [- c \pm (c^2 - 4\omega^2)^{1/2}]/2.$$

If $c^2 > 4\omega^2$ then λ_1 and λ_2 are both negative so $(n\pi, 0)$ is a stable node, while if $c^2 < 4\omega^2$ then λ_1 and λ_2 are complex conjugates with negative real part, so $(n\pi, 0)$ is a stable spiral point.

Problems 10-14 call for us to find and classify the critical points of the first order system

$$x' = y$$
$$y' = -f(x, y)$$

that corresponds to the given equation

$$x'' + f(x, x') = 0.$$

We give here only the answers.

10. The critical points $(\pm 2, 0)$ are unstable saddle points, while $(0, 0)$ is a stable center of the linearized system.

11. The critical points $(\pm 2, 0)$ are unstable saddle points, while $(0, 0)$ is a stable spiral point.

12. The critical points $(\pm 2, 0)$ are stable centers of the linearized system, while $(0, 0)$ is an unstable saddle point.

13. The origin (0,0) is a stable center of the linearized system, while (-4,0) is an unstable saddle point.

14. The critical points (±2,0) and (0,0) are stable centers of the linearized system, while the critical points (±1,0) are unstable saddle points

CHAPTER 8

FOURIER SERIES AND SEPARATION OF VARIABLES

SECTION 8.1

PERIODIC FUNCTIONS AND TRIGONOMETRIC SERIES

1.   Smallest period $P = 2\pi/3$

2.   Smallest period $P = 1$

3.   Smallest period $P = 4\pi/3$

4.   Smallest period $P = 6$

5.   Smallest period $P = \pi$

6.   Smallest period $P = 1/2$

7.   Not periodic

8.   Not periodic

9.   Smallest period $P = \pi$

10. Smallest period $P = \pi/3$

11. With $f(t) = 1$ the integral formulas in (16) and (17) give
     $a_0 = 2$ and $a_n = b_n = 0$ for $n \geq 0$. Thus the Fourier
     series of f is the single term series $f(t) = 1$.

12. $-\dfrac{12}{\pi}\left[\dfrac{\sin t}{1}+\dfrac{\sin 3t}{3}+\dfrac{\sin 5t}{5}+\cdots\right]$

13. $\dfrac{1}{2}+\dfrac{2}{\pi}\left[\dfrac{\sin t}{1}+\dfrac{\sin 3t}{3}+\dfrac{\sin 5t}{5}+\cdots\right]$

14. $\dfrac{1}{2}-\dfrac{10}{\pi}\left[\dfrac{\sin t}{1}+\dfrac{\sin 3t}{3}+\dfrac{\sin 5t}{5}+\cdots\right]$

15. $2\left[\dfrac{\sin t}{1}+\dfrac{\sin 2t}{2}+\dfrac{\sin 3t}{3}+\cdots\right]$

16. $\pi-2\left[\dfrac{\sin t}{1}+\dfrac{\sin 2t}{2}+\dfrac{\sin 3t}{3}+\cdots\right]$

17. $\dfrac{\pi}{2}-\dfrac{4}{\pi}\left[\dfrac{\cos t}{1}+\dfrac{\cos 3t}{9}+\dfrac{\cos 5t}{25}+\cdots\right]$

18. $\dfrac{\pi}{2}+\dfrac{4}{\pi}\left[\dfrac{\cos t}{1}+\dfrac{\cos 3t}{9}+\dfrac{\cos 5t}{25}+\cdots\right]$

19. $\dfrac{\pi}{4}+\dfrac{2}{\pi}\left[\dfrac{\cos t}{1}+\dfrac{\cos 3t}{9}+\dfrac{\cos 5t}{25}+\cdots\right]$

$\qquad -\left[\dfrac{\sin t}{1}+\dfrac{\sin 2t}{2}+\dfrac{\sin 3t}{3}+\cdots\right]$

20. $\dfrac{1}{2}+\dfrac{2}{\pi}\left[\dfrac{\cos t}{1}-\dfrac{\cos 3t}{3}+\dfrac{\cos 5t}{5}-\cdots\right]$

21. $\dfrac{\pi^2}{3}-4\left[\dfrac{\cos t}{1}-\dfrac{\cos 2t}{4}+\dfrac{\cos 3t}{9}-\cdots\right]$

22. $\dfrac{4\pi^2}{3}+4\left[\dfrac{\cos t}{1}+\dfrac{\cos 2t}{4}+\dfrac{\cos 3t}{9}+\cdots\right]$

$\qquad -4\pi\left[\dfrac{\sin t}{1}+\dfrac{\sin 2t}{2}+\dfrac{\sin 3t}{3}+\cdots\right]$

23. $\dfrac{\pi^2}{6} - 2\left[\dfrac{\cos t}{1} - \dfrac{\cos 2t}{4} + \dfrac{\cos 3t}{9} - \cdots\right]$

$+ \pi\left[\dfrac{\sin t}{1} - \dfrac{\sin 2t}{2} + \dfrac{\sin 3t}{3} - \cdots\right]$

$- \dfrac{2}{\pi}\left[\dfrac{\sin t}{1} + \dfrac{\sin 3t}{27} + \dfrac{\sin 5t}{125} + \cdots\right]$

The trigonometric identities

$$2\cos A\cos B = \cos(A + B) + \cos(A - B)$$
$$2\sin A\cos B = \sin(A + B) + \sin(A - B)$$
$$2\sin A\sin B = \cos(A - B) - \cos(A + B)$$

are needed to evaluate the integrals that appear in Problems 24-26.

24. $b_n = 0$ for $n = 1,2,3,\cdots$, $a_n = 0$ for n odd, and $a_n = -4/\pi(n^2 - 1)$ if $n = 0,2,4,\cdots$; the series is

$$\dfrac{2}{\pi} - \dfrac{4}{\pi}\left[\dfrac{\cos 2t}{3} + \dfrac{\cos 4t}{15} + \dfrac{\cos 6t}{35} + \cdots\right]$$

25. In order to evaluate the coefficient integrals in (16) and (17) we would need the trigonometric identity

$$\cos^2 2t = (1/2)(1 + \cos 4t)$$

which, however, tells us in advance that the coefficients in the Fourier series of $f(t) = \cos^2 2t$ are given by $a_0 = 1$, $a_4 = 1/2$, $a_n = 0$ otherwise and $b_n = 0$ for all $n \geq 1$.

26. $\dfrac{1}{2}\sin t + \dfrac{1}{\pi} - \dfrac{2}{\pi}\left[\dfrac{\cos 2t}{3} + \dfrac{\cos 4t}{15} + \dfrac{\cos 6t}{35} + \cdots\right]$

Note that $f(t) = (\sin t + |\sin t|)/2$, so this answer agrees with the answer to Problem 24.

GENERAL FOURIER SERIES AND CONVERGENCE

1. $\frac{8}{\pi} \left[\sin \frac{\pi t}{3} + \frac{1}{3} \sin \frac{3\pi t}{3} + \frac{1}{5} \sin \frac{5\pi t}{3} + \cdots \right]$

2. $\frac{1}{2} + \frac{2}{\pi} \left[\sin \frac{\pi t}{5} + \frac{1}{3} \sin \frac{3\pi t}{5} + \frac{1}{5} \sin \frac{5\pi t}{5} + \cdots \right]$

3. $\frac{1}{2} + \frac{6}{\pi} \left[\sin \frac{t}{2} + \frac{1}{3} \sin \frac{3t}{2} + \frac{1}{5} \sin \frac{5t}{2} + \cdots \right]$

4. $\frac{4}{\pi} \left[\sin \frac{\pi t}{2} - \frac{1}{2} \sin \frac{2\pi t}{2} + \frac{1}{3} \sin \frac{3\pi t}{2} - \cdots \right]$

5. $4 \left[\sin \frac{t}{2} - \frac{1}{2} \sin \frac{2t}{2} + \frac{1}{3} \sin \frac{3t}{2} - \cdots \right]$

6. $\frac{3}{2} - \frac{3}{\pi} \left[\sin \frac{2\pi t}{3} + \frac{1}{2} \sin \frac{4\pi t}{3} + \frac{1}{3} \sin \frac{6\pi t}{3} + \cdots \right]$

7. $\frac{1}{2} - \frac{4}{\pi^2} \left[\cos \pi t + \frac{1}{9} \cos 3\pi t + \frac{1}{25} \cos 5\pi t + \cdots \right]$

8. $L = 3/2$ and $a_0 = 2/3$. We find that

$$a_n = \frac{1}{n\pi} \left[\sin \frac{4n\pi}{3} - \sin \frac{2n\pi}{3} \right],$$

$$b_n = \frac{1}{n\pi} \left[\cos \frac{2n\pi}{3} - \cos \frac{4n\pi}{3} \right].$$

Analyzing separately the cases $n = 3k$, $n = 3k + 1$, and $n = 3k + 2$, we find that $a_{3k} = 0$, $a_{3k+1} = -\sqrt{3}/\pi n$, $a_{3k+2} = +\sqrt{3}/\pi n$, and that $b_n = 0$ for all n. Hence the Fourier series of $f(t)$ is

$$\frac{1}{3} - \frac{\sqrt{3}}{\pi} \left[\cos \frac{2\pi t}{3} - \frac{1}{2} \cos \frac{4\pi t}{3} + \frac{1}{4} \cos \frac{8\pi t}{3} - \frac{1}{5} \cos \frac{10\pi t}{3} + \cdots\right]$$

9. $\frac{1}{3} - \frac{4}{\pi^2} \left[\cos \pi t - \frac{1}{4} \cos 2\pi t + \frac{1}{9} \cos 3\pi t - \cdots\right]$

10. $\frac{2}{3} - \frac{8}{\pi^2} \left[\cos \frac{\pi t}{2} - \frac{1}{4} \cos \frac{2\pi t}{2} + \frac{1}{9} \cos \frac{3\pi t}{2} - \cdots\right]$

$\qquad + \frac{4}{\pi} \left[\sin \frac{\pi t}{2} - \frac{1}{2} \sin \frac{2\pi t}{2} + \frac{1}{3} \sin \frac{3\pi t}{2} - \cdots\right]$

$\qquad - \frac{16}{\pi^3} \left[\sin \frac{\pi t}{2} + \frac{1}{27} \sin \frac{3\pi t}{2} + \frac{1}{125} \sin \frac{5\pi t}{2} + \cdots\right]$

To calculate the Fourier coefficients in Problems 11-14 we use the trigonometric identities for $\sin A \cos B$ and $\sin A \sin B$ that are listed above in Section 8.1.

11. $a_0 = 4/\pi$, $a_n = (4/\pi)(-1)^{n+1}/(4n^2 - 1)$ for $n \geq 1$, and $b_n = 0$ for $n \geq 1$, so the Fourier series is

$$\frac{2}{\pi} + \frac{4}{\pi} \left[\frac{1}{3} \cos \pi t - \frac{1}{15} \cos 2\pi t + \frac{1}{35} \cos 3\pi t - \cdots\right]$$

12. $L = 1/2$, $a_0 = 4/\pi$ and $a_n = -4/\pi(4n^2 - 1)$, $b_n = 0$ for $n = 1,2,3,\cdots$, so the period 1 Fourier series of $f(t) = \sin \pi t$ is

$$\frac{2}{\pi} - \frac{4}{\pi} \left[\frac{1}{3} \cos 2\pi t + \frac{1}{15} \cos 4\pi t + \frac{1}{35} \cos 6\pi t + \cdots\right]$$

13. We find that $a_0 = 2/\pi$, $a_1 = 0$, and

$$a_n = [(-1)^{n+1} - 1]/[\pi(n^2 - 1)]$$

for $n \geq 2$, while $b_1 = 1/2$ and $b_n = 0$ for $n \geq 2$. Hence the Fourier series of $f(t)$ is

$$\frac{1}{\pi} + \frac{1}{2\pi} \sin \pi t$$

$$- \frac{2}{\pi} \left[\frac{1}{3} \cos 2\pi t + \frac{1}{15} \cos 4\pi t + \frac{1}{35} \cos 6\pi t + \cdots \right]$$

14. $L = 2\pi$ and $a_0 = 0$. If $n \neq 2$ then $a_n = 0$ if n is even, $a_n = -4/\pi(n^2 - 4)$ if n is odd, and $b_n = 0$. Working separately with $n = 2$ we calculate $a_2 = 0$, $b_2 = 1/2$. Hence the desired Fourier series is

$$\frac{1}{2} \sin t + \frac{4}{\pi} \left[\frac{1}{3} \cos \frac{t}{2} - \frac{1}{5} \cos \frac{3t}{2} - \frac{1}{21} \cos \frac{5t}{2} - \cdots \right]$$

15. (b) Substitute $t = 0$ in the Fourier series of part (a) to obtain the series in (16). Note that

$$f(0) = [f(0-) + f(0+)]/2 = [(2\pi)^2 + (0)^2]/2 = 2\pi^2.$$

Substitute $t = \pi$ and $f(\pi) = \pi^2$ to obtain the series in (17).

16. (b) Substitute $t = 0$.

17. (b) Substitute $t = 1/2$.

The most efficient approach to Problems 18 and 20 is to derive first the expansions

$$t = \pi - 2 \left[\sin t + \frac{\sin 2t}{2} + \frac{\sin 3t}{3} + \cdots \right],$$

$$t^2 = \frac{4\pi^2}{3} + 4 \left[\cos t + \frac{\cos 2t}{4} + \frac{\cos 3t}{9} + \cdots \right]$$

$$- 4\pi \left[\sin t + \frac{\sin 2t}{2} + \frac{\sin 3t}{3} + \cdots \right]$$

for $0 < t < 2\pi$, as the Fourier series of the functions $f(t)$ and $g(t)$ of period 2π defined for $0 < t < 2\pi$ by $f(t) = t$

and $g(t) = t^2$. The first series above yields the series in Problem 18, and a combination of the two yields the series in Problem 20.

The expansions in Problems 19 and 21 are valid on the interval $-\pi < t < \pi$ rather than the interval $0 < t < 2\pi$. When we calculate the Fourier series of the functions $f(t)$ and $g(t)$ of period 2π defined for $-\pi < t < \pi$ by $f(t) = t$ and $g(t) = t^2$, we find that

$$t = 2 \left[\sin t - \frac{\sin 2t}{2} + \frac{\sin 3t}{3} - \cdots \right].$$

$$t^2 = \frac{\pi^2}{3} - 4 \left[\cos t - \frac{\cos 2t}{4} + \frac{\cos 3t}{9} - \cdots \right]$$

if $-\pi < t < \pi$.

24. (b) When we substitute $t = 0$ in the series of part (a) we get

$$8\pi^4 = 16\pi^4/5 + 32\pi^2 \sum(1/n^2) - 48 \sum(1/n^4).$$

After substituting $\sum(1/n^2) = \pi^2/6$ we solve this equation for $\sum(1/n^4) = \pi^4/90$. Similarly, we find that $\sum(-1)^{n+1}/n^4 = 7\pi^4/720$ by substituting $t = \pi$ in the series of part (a). Finally, addition of the first two series in part (b) yields the third one.

25. Now we want to sum the series

$$1 - \frac{1}{3^3} + \frac{1}{5^3} - \frac{1}{7^3} + \cdots.$$

Having used a Fourier series of t^4 in Problem 24 to evaluate $\sum(1/n^4)$, it is natural to look at a Fourier series of t^3. Let $f(t)$ be the period 2π function with $f(t) = t^3$ if $-\pi < t < \pi$. We use Formulas (22)-(25) in

Section 7.1 to calculate the Fourier coefficients of $f(t)$, and find that

$$t^3 = 2\pi^2 \sum_1^\infty (-1)^{n+1} \frac{\sin nt}{n} - 12 \sum_1^\infty (-1)^{n+1} \frac{\sin nt}{n^3} .$$

If we substitute $t = \pi/2$ and use Leibniz's series $\sum(-1)^{n+1}/n = \pi/4$ of Problem 17 we find that

$$1 - \frac{1}{3^3} + \frac{1}{5^3} - \frac{1}{7^3} + \cdots = \frac{\pi^3}{32} .$$

There is no value of t whose substitution in the Fourier series of $f(t) = t^3$ yields the series $\sum(1/n^3)$ containing the reciprocal cubes of both the odd and even integers. Indeed, the summation in "closed form" of the series

$$1 + \frac{1}{2^3} + \frac{1}{3^3} + \frac{1}{4^3} + \cdots$$

is a problem that has challenged many fine mathematicians since the time of Euler. Only recently (by R. Apery in 1978) has it been shown that this sum is an irrational number. For a delightful account of this work, see the article "A Proof that Euler Missed ... An Informal Report" by Alfred van der Poorten in the *The Mathematical Intelligencer*, Volume 1 (1979), pages 195-203.

SECTION 8.3

EVEN-ODD FUNCTIONS AND TERMWISE DIFFERENTIATION

1. *Cosine series*: $a_0 = 2$; the coefficient of $\cos n\pi t/L$ is $a_n = 0$ for $n \geq 1$.

    Sine series: the coefficient of $\sin n\pi t/L$ is $b_n = 4/n\pi$

for n odd, $b_n = 0$ for n even.

2. *Cosine series:* $a_0 = 1$; the coefficient of cos nπt is a_n $= -4/n^2\pi^2$ for n odd, $a_n = 0$ for n even.

Sine series: The coefficient of sin nπt is $b_n = 2/n\pi$ for $n = 1, 2, 3, \cdots$.

3. *Cosine series:* $a_0 = 0$; the coefficient of cos nπt/2 is

$$a_n = [1 - (-1)^n](4/n^2\pi^2)$$

for $n \geq 1$, so $a_n = 8/n^2\pi^2$ for n odd, $a_n = 0$ for n even.

Sine series: The coefficient of sin nπt/2 is

$$b_n = [1 + (-1)^n](2/n\pi)$$

for $n \geq 1$, so $b_n = 4/n\pi$ for n even, $b_n = 0$ for n odd.

4. *Cosine series:* $a_0 = 1$; the coefficient of cos(nπt/2) is $a_n = 0$ for n odd, $a_n = 0$ if $n = 4k$, $a_n = -16/n^2\pi^2$ if $n = 4k + 2$.

Sine series The coefficient of sin(nπt/2) is given by $b_n = 0$ for n even, $b_n = 8/n^2\pi^2$ if $n = 4k + 1$, and $b_n = -8/n^2\pi^2$ if $n = 4k + 3$.

5. *Cosine series:* $a_0 = 2/3$; the coefficient of cos nπt/3 is $a_n = -(2\sqrt{3})/n\pi$ if n is of the form $6k + 2$, $a_n =$ $+(2\sqrt{3})/n\pi$ if n is of the form $6k + 4$, and $a_n = 0$ otherwise. Thus the Fourier cosine series of f(t) is

$$\frac{1}{3} - \frac{2\sqrt{3}}{\pi} \left[\frac{1}{2} \cos \frac{2\pi t}{3} - \frac{1}{4} \cos \frac{4\pi t}{3} + \frac{1}{8} \cos \frac{8\pi t}{3} \right.$$

$$- \frac{1}{10} \cos \frac{10\pi t}{3} + \cdots]$$

Sine series: The coefficient of $\sin n\pi t/3$ is $b_n = 0$ if n is even, $b_n = -4/n\pi$ if n is of the form $6k + 3$, and $b_n = +2/n\pi$ if either $n = 6k + 1$ or $n = 6k + 5$.

6. *Cosine series:* $a_0 = 2L^2/3$; the coefficient of $\cos(n\pi t/L)$ is given by $a_n = 4L^2(-1)^n/n^2\pi^2$ for $n = 1, 2, 3, \cdots$.

   Sine series: The coefficient of $\sin(n\pi t/L)$ is $b_n = 2L^2(n^2\pi^2 - 4)/n^3\pi^3$ for n odd, $b_n = -2L^2/n\pi$ for n even.

7. *Cosine series:* $a_0 = \pi^2/3$; the coefficient of $\cos nt$ is $a_n = -[1 + (-1)^n](2/n^2)$ for $n \geq 1$.

   Sine series: The coefficient of $\sin nt$ is $b_n = [1 - (-1)^n](4/\pi n^3)$ for $n \geq 1$.

8. *Cosine series:* $a_0 = 1/3$; the coefficient of $\cos n\pi t$ is given by $a_n = -2(-1)^n/n^2\pi^2$ for $n = 1, 2, 3, \cdots$.

   Sine series: The coefficient of $\sin n\pi t$ is $b_n = 8/n^3\pi^3$ for n odd, $b_n = 0$ for n even. Thus

$$t - t^2 = \frac{8}{\pi^3} \left[\sin \pi t + \frac{\sin 3\pi t}{27} + \frac{\sin 5\pi t}{125} + \cdots \right]$$

   for $0 < t < 1$. Note that $t = 1/2$ yields the summation of Problem 25 in Section 8.2.

9. *Cosine series:* $a_n = 4/\pi$ and $a_1 = 0$; if $n \geq 2$, then the coefficient of $\cos nt$ is $a_n = -[1 + (-1)^n]/[2/\pi(n^2 - 1)]$.

   Sine series: $b_1 = 1$ and $b_n = 0$ for $n \geq 2$.

10. *Cosine series:* $a_0 = 2/\pi$; the coefficient of $\cos(nt/2)$ is

$a_n = -4/\pi(n^2 - 4)$ if n is odd, $a_n = -8/\pi(n^2 - 4)$ if $n = 4k$, and $a_n = 0$ if $n = 4k + 2$.

Sine series : The coefficient of $\sin(nt/2)$ is $b_n = 0$ if n is even, $b_n = -4/\pi(n^2 - 4)$ if $n = 4k + 1$, and $b_n = 4/\pi(n^2 - 4)$ if $n = 4k + 3$.

11. $x(t) = -\dfrac{4}{\pi} \displaystyle\sum_{n \text{ odd}} \dfrac{1}{n(n^2 - 2)} \sin nt$

12. Because of the initial conditions $x(0) = x(\pi) = 0$, we try a sine series solution of the form

$$x(t) = \sum b_n \sin nt.$$

Then

$$x''(t) - 4x(t) = -\sum (n^2 + 4)b_n \sin nt,$$

and

$$1 = (4/\pi)[\sin t + (1/3)\sin 3t + (1/5)\sin 5t + \cdots]$$

by Example 1 in Section 8.1. We therefore choose

$$b_n = 0 \quad \text{if} \quad n \text{ is even,}$$
$$b_n = -4/\pi n(n^2 + 4) \quad \text{if} \quad n \text{ is odd}$$

to get the formal Fourier series solution

$$x(t) = -\dfrac{4}{\pi} \left[\dfrac{\sin t}{5} + \dfrac{\sin 3t}{39} + \dfrac{\sin 5t}{145} + \cdots \right].$$

13. $x(t) = \displaystyle\sum_{n=1}^{\infty} \dfrac{2(-1)^n}{n\pi(n^2\pi^2 - 1)} \sin n\pi t$

14. To solve the endpoint value problem

$$x'' + 2x = t, \quad x(0) = x(2) = 0,$$

we try a sine series solution of the form

$$x(t) = \sum b_n \sin(n\pi t/2).$$

Then

$$x''(t) + 2x(t) = \sum [2 - n^2\pi^2/4]b_n \sin(n\pi t/2),$$

and Example 1 in this section gives

$$t = (4/\pi) \sum [(-1)^{n+1}/n] \sin(n\pi t/2).$$

When we equate the coefficients of $\sin(n\pi t/2)$ in these two series, we find that we must choose

$$b_n = 16(-1)^{n+1}/\pi n(8 - n^2\pi^2)$$

to get a formal Fourier series solution.

15. $$x(t) = \frac{\pi}{2} + \frac{2}{\pi} \sum_{n=1}^{\infty} \frac{1 - (-1)^n}{n^2(n^2 - 2)} \cos nt$$

16. (b) The point is simply that the series in (31) is the Fourier sine series of the period 2 function defined by $f(t) = t - (\sin 2t)/(\sin 2)$ for $0 < t < 1$.

17. *Suggestion*: Substitute $u = -t$ in the left-hand integral.

18. The termwise derivative of the given Fourier series is

$$- (4/\pi) \sum (\sin n\pi t)/n - 4 \sum \cos n\pi t.$$

But the series $\sum \cos n\pi t$ diverges at $t = 0$ (for instance). Hence the derived series does not converge to any function at all, let alone to $f'(t)$.

22. We want to calculate the coefficients in the Fourier cosine series

$$G(t) = \frac{a_0}{2} + \sum_{1}^{\infty} a_n \cos \frac{n\pi t}{2L},$$

which agrees with $f(t)$ if $0 < t < L$. Then

$$a_n = \frac{2}{2L} \int_0^L f(t) \cos \frac{n\pi t}{2L}\, dt - \frac{2}{2L} \int_L^{2L} f(2L - t) \cos \frac{n\pi t}{2L}\, dt.$$

The substitution $u = 2L - t$ yields

$$a_n = \frac{1}{L} \int_0^L f(t) \cos \frac{n\pi t}{2L}\, dt + \frac{1}{L} \int_L^0 f(u) \cos \frac{n\pi(2L - u)}{2L}\, du$$

$$= \frac{1}{L} \int_0^L f(t) \cos \frac{n\pi t}{2L}\, dt - \frac{(-1)^n}{L} \int_0^L f(u) \cos \frac{n\pi u}{2L}\, du.$$

Now it is clear that

$$a_n = \frac{2}{L} \int_0^L f(t) \cos \frac{n\pi t}{2L}\, dt$$

if n is odd, while $a_n = 0$ if n is even (including $n = 0$).

24. We want a solution of the form

$$x(t) = \sum B_n \sin(nt/2)$$

with $B_n = 0$ for n even, because each term of such a series will satisfy the endpoint conditions $x(0) = x'(\pi) = 0$. Then

$$x''(t) - x(t) = -\sum (1 + n^2/4) B_n \sin(nt/2),$$

and by Problem 23 we have

$$t = \sum b_n \sin(nt/2)$$

where $b_n = 0$ for n even, and otherwise

$$b_n = 8(-1)^{(n-1)/2}/\pi n^2.$$

When we equate these two series we get

$$B_n = 32(-1)^{(n+1)/2}/\pi n^2(n^2 + 4) \quad \text{for} \quad n \quad \text{odd.}$$

SECTION 8.4

APPLICATIONS OF FOURIER SERIES

1. $\quad x_{sp}(t) = \dfrac{12}{\pi} \displaystyle\sum_{n \text{ odd}} \dfrac{\sin nt}{n(5 - n^2)}$

2. First we calculate the Fourier cosine series of the even period 4 function $F(t)$, and find that

$$F(t) = \sum a_n \cos(n\pi t/2)$$

where $a_n = (12/n\pi)\sin(n\pi/2)$, so $a_n = 0$ for n even and $a_n = [12(-1)^{(n-1)/2}]/n\pi$ for n odd. Then we substitute this series and the steady periodic trial solution

$$x(t) = \sum A_n \cos(n\pi t/2)$$

in $x'' + 10x = F(t)$, and find that we must choose $A_n = 4a_n/(40 - n^2\pi^2)$. Thus

$$A_n = [48(-1)^{(n-1)/2}]/[n\pi(40 - n^2\pi^2)]$$

for n odd, and $A_n = 0$ for n even.

3. $\quad x_{sp}(t) = \displaystyle\sum_{n=1}^{\infty} \dfrac{(-1)^n \sin nt}{n(n^2 - 3)}$

4. The Fourier cosine series of the period 4 function $F(t)$ is

$$F(t) = a_0 + \sum a_n \cos(n\pi t/2)$$

where $a_0 = 2$, $a_n = 0$ for n even $(n > 0)$, and $a_n = -16/n^2\pi^2$ for n odd. When we substitute this series and the steady periodic trial solution

$$x(t) = A_0 + \sum A_n \cos(n\pi t/2)$$

in the differential equation $x'' + 4x = F(t)$, we find that $A_0 = a_0/4 = 1/2$ and

$$(4 - n^2\pi^2/4)A_n = a_n$$

for $n > 0$. Therefore we choose

$$A_n = 64/n^2\pi^2(n^2\pi^2 - 16)$$

for n odd, and $A_n = 0$ for n even.

5. $$x_{sp}(t) = \frac{8}{\pi^3} \sum_{n \text{ odd}} \frac{\sin n\pi t}{n^3(10 - n^2\pi^2)}$$

6.   The period 2π Fourier cosine series of $F(t) = \sin t$ is

$$F(t) = a_0 + \sum a_n \cos nt$$

where $a_0 = 4/\pi$, $a_n = 0$ for n odd, and $a_n = -4/\pi(n^2 - 1)$ for n even $(n > 0)$. Substitution of this series and the steady periodic trial solution

$$x(t) = A_0 + \sum A_n \cos nt$$

in $x'' + 2x = F(t)$ yields $A_0 = a_0/2 = 2/\pi$ and $(2 - n^2)A_n = a_n$ for $n > 0$. Hence we choose

$$A_n = 4/[\pi(n^2 - 1)(n^2 - 2)]$$

for n even and $A_n = 0$ for n odd.

In Problems 7-12 we are dealing with the equation

$$mx'' + kx = F(t)$$

where $F(t)$ is the external periodic force. The natural frequency is $\omega_0 = \sqrt{(k/m)}$. If the Fourier series of $F(t)$ contains a term of the form $\cos(N\pi t/L)$ or $\sin(N\pi t/L)$ with $\omega_0 = N\pi/L$, then pure resonance occurs. Otherwise, it does not.

7.   The natural frequency is $\omega_0 = 3$, and $F(t) = \sum b_n \sin nt$ with $b_n = 4/n\pi$ for n odd, $b_n = 0$ for n even. Thus the Fourier series of $F(t)$ contains a $\sin 3t$ term, so resonance does occur.

8.   The natural frequency is $\omega_0 = \sqrt{5}$, and $F(t) = \sum b_n \sin n\pi t$. Since $n\pi \neq \sqrt{5}$ for any integer n, pure resonance does not occur.

9.   The natural frequency is $\omega_0 = 2$, and $F(t) = \sum b_n \sin nt$ with $b_n = 12/n\pi$ for n odd, $b_n = 0$ for n even. Because the $\sin 2t$ term is missing from the Fourier series of $F(t)$, resonance will not occur.

10. The natural frequency is $\omega_0 = 2\pi$. From Equation (16) in Section 8.3 we see that the Fourier series of $F(t)$ contains a $\sin 2\pi t$ term. Hence pure resonance occurs.

11. The natural frequency is $\omega_0 = 4$. From Equation (15) in Section 8.3 we see that

$$F(t) = \pi/2 - (4/\pi)[(\cos t) + (\cos 3t)/9 + (\cos 5t)/25 + \cdots].$$

Because the $\cos 4t$ term is missing, we see that resonance will not occur.

12. The natural frequency is $\omega_0 = 5$, and the Fourier series of $F(t)$ is of the form

$$F(t) = \sum b_n \sin nt.$$

We calculate b_5, and find that $b_5 = 8/125\pi$. Thus the term sin 5t is present in F(t), and so pure resonance occurs.

Problems 13-18 are based on Equations (14)-(16) in the text, according to which the steady periodic solution of

$$mx'' + cx' + kx = \sum B_n \sin(n\pi t/L)$$

is given by

$$x_{sp}(t) = \sum b_n \sin(\omega_n t - \alpha_n)$$

where

$$\omega_n = n\pi/L, \quad \alpha_n = \tan^{-1}(c\omega_n/[k - m(\omega_n)^2]) \quad \text{in } [0,\pi],$$

$$b_n = B_n/([k - m(\omega_n)^2]^2 + (c\omega_n)^2)^{1/2}.$$

The following program was written to calculate the coefficients $\{b_n\}$ for Problem 13. The values of m,c,k,L are input, and only line 300 must be changed for problems 14-18.

```
100 REM--Damped Forced Oscillations Program
110 REM--(Problem 13, Section 7.4)
120 REM--Input parameters:
130       DEFINT N
140       INPUT "Mass m"; M
150       INPUT "Dashpot c"; C
160       INPUT "Spring k"; K
170       INPUT "Half-period L"; L
180       PI = 3.141593  :  EPS = .00001
185       PRINT "N        B              OMEGA          ALPHA"
190 REM--Compute displacement coefficients:
200       FOR N = 1 TO 9
210            GOSUB 300  :   REM--Get BN
220            W = N*PI/L :   REM-- W = OMEGA
225            D = K - M*W*W
230            IF ABS(D) < EPS THEN ALPHA = PI/2
                    ELSE ALPHA = ATN(C*W/D)
240            IF ALPHA < 0 THEN ALPHA = PI + ALPHA
250            X = B/SQR(D*D + C*C*W*W)
260            PRINT USING
                  "#   #.#####     ##.#####    #.#####";
                  N, X, W, ALPHA
270       NEXT N
280       STOP
290 REM--Compute force coefficients:
```

```
300       IF N\2 = N/2 THEN B = 0 ELSE B = 12/(PI*N)
310       RETURN
320       END
```

13. $B_n = 12/\pi n$ for n odd, $B_n = 0$ for n even

$$x_{sp}(t) \approx 1.2725 \sin(t - 0.0333) + 0.2542 \sin(3t - 3.0817)$$
$$+ 0.0364 \sin(5t - 3.1178) + \cdots$$

14. $B_n = 4(-1)^{n+1}/n$ for $n = 1,2,3,\cdots$

$$x_{sp}(t) \approx 0.2500 \sin(t - 0.0063) - 0.2000 \sin(2t - 0.0200)$$
$$+ 4.444 \sin(3t - 1.5708) - 0.0714 \sin(4t - 3.1130)$$
$$+ \cdots$$

Note the dominance of the n = 3 term.

15. $B_n = 8/n^3\pi^3$ for n odd, $B_n = 0$ for n even

$$x_{sp}(t) \approx 0.08150 \sin(\pi t - 1.44692)$$
$$+ 0.00004 \sin(3\pi t - 3.10176) + \cdots$$

16. $F(t) = A_0 + \sum A_n \cos(n\pi t/2)$ where $A_0 = 2$, $A_n = -16/\pi^2 n^2$
for n odd, $A_n = 0$ for n even.

$$x_{sp}(t) \approx 0.5000 + 1.0577 \cos(\pi t/2 - 0.0103)$$
$$- 0.0099 \cos(3\pi t/2 - 3.1390)$$
$$- 0.0011 \cos(5\pi t/2 - 3.1402) \cdots$$

17. $B_n = 60/n\pi$ for n odd, $B_n = 0$ for n even

$$x_{sp}(t) \approx 0.5687 \sin(\pi t - 0.0562) + 0.4271 \sin(3\pi t - 0.3891)$$
$$+ 0.1396 \sin(5\pi t - 2.7899) + 0.0318 \sin(7\pi t - 2.9874)$$
$$+ \cdots$$

$$x_{sp}(5) \approx 0.248 \text{ ft} \approx 2.98 \text{ in.}$$

18. $B_n = (4/\pi n^2)\sin(n\pi/2)$

$x_{sp}(t) \approx 0.0531 \sin(t - 0.0004) - 0.0088 \sin(3t - 0.0019)$
$\qquad + 1.0186 \sin(5t - 1.5708) - 0.0011 \sin(7t - 3.1387)$
$\qquad + \cdots$

Note the dominance of the $n = 5$ term.

SECTION 8.5

HEAT CONDUCTION AND SEPARATION OF VARIABLES

1. From Equation (31) in the text, with $L = \pi$ and $k = 3$, we get

$$u(x,t) = \sum_{n=1}^{\infty} b_n \exp(-3n^2 t)\sin nx.$$

With $b_2 = 4$ and $b_n = 0$ otherwise we get the solution

$$u(x,t) = 4e^{-12t}\sin 2x.$$

2. From Equation (40) with $k = 10$ and $L = 5$ we get

$$u(x,t) = a_0/2 + \sum a_n \exp(-10n^2\pi^2 t/25)\cos(n\pi x/5).$$

With $a_0 = 14$ and $a_n = 0$ for $n > 0$ we get the solution $u(x,t) = 7$ (constant).

3. With $L = 1$ and $k = 2$ in Equation (31), we take $b_1 = 5$, $b_3 = -1/5$, and $b_n = 0$ otherwise. The result is the solution

$$u(x,t) = 5 \exp(-2\pi^2 t)\sin \pi x - (1/5)\exp(-18\pi^2 t)\sin 3\pi x.$$

4. From Equation (31) with $k = 1$ and $L = \pi$ we get

$$u(x,t) = \sum b_n \exp(-n^2 t) \sin nx.$$

But the sin A cos B identity yields

$$4 \sin 4x \cos 2x = 2 \sin 2x + 2 \sin 6x.$$

Hence we choose $b_2 = b_6 = 2$ and $b_n = 0$ for $n \neq 2,6$. Thus

$$u(x,t) = 2e^{-4t} \sin 2x + 2e^{-36t} \sin 6x.$$

5. $u(x,t) = 4 \exp(-8\pi^2 t/9) \cos(2\pi x/3) - 2\exp(-32\pi^2 t/9) \cos(4\pi x/3)$

6. From Equation (31) with $k = 1/2$ and $L = 1$ we get

$$u(x,t) = \sum b_n \exp(-n^2 \pi^2 t/2) \sin n\pi x.$$

Trigonometric identities yield

$$\begin{aligned}
4 \sin \pi x \cos^3 \pi x &= (2 \sin \pi x \cos \pi x)(2 \cos^2 \pi x) \\
&= (\sin 2\pi x)(1 + \cos 2\pi x) \\
&= \sin 2\pi x + (1/2)\sin 4\pi x.
\end{aligned}$$

Hence we choose $b_2 = 1$, $b_4 = 1/2$, and $b_n = 0$ otherwise to get

$$u(x,t) = \exp(-2\pi^2 t)\sin 2\pi x + (1/2)\exp(-8\pi^2 t)\sin 4\pi x.$$

7. The identity $\cos^2 2\pi x = (1 + \cos 4\pi x)/2$ yields

$$u(x,t) = 1/2 + (1/2)\exp(-16\pi^2 t/3)\cos 4\pi x.$$

8. From Equation (40) with $k = 1$ and $L = 2$ we get

$$u(x,t) = a_0/2 + \sum a_n \exp(-n^2 \pi^2 t/4)\cos n\pi x/2.$$

But

$$10 \cos \pi x \cos 3\pi x = 5 \cos 2\pi x + 5 \cos 4\pi x.$$

Hence we choose $b_4 = b_8 = 5$ and $b_n = 0$ otherwise to get

$$u(x,t) = 5 \exp(-4\pi^2 t)\cos 2\pi x + 5 \exp(-16\pi^2 t)\cos 4\pi x.$$

9. $$u(x,t) = \frac{100}{\pi} \sum_{n \text{ odd}} \frac{1}{n} \exp\left(\frac{-n^2\pi^2 t}{250}\right)\sin \frac{n\pi x}{5}$$

10. From Equation (31) with $k = 1/5$ and $L = 10$ we get

    $$u(x,t) = \sum b_n \exp(-n^2\pi^2 t/500)\sin n\pi x/10.$$

    On the basis of Equation (16) in Section 8.3 we choose

    $$b_n = 80(-1)^{n+1}/\pi n \quad \text{for} \quad n = 1,2,3,\cdots.$$

11. $$u(x,t) = 20 - \frac{160}{\pi^2} \sum_{n \text{ odd}} \frac{1}{n^2} \exp\left(\frac{-n^2\pi^2 t}{500}\right)\cos \frac{n\pi x}{10}$$

12. From Equation (31) with $k = 1$ and $L = 100$ we get

    $$u(x,t) = \sum b_n \exp(-n^2\pi^2 t/10000)\sin n\pi x/100.$$

    The $\{b_n\}$ should be the Fourier sine coefficients of $f(x)$ = $x(100 - x)$ on $[0,100]$. Hence we choose $b_n = 80000/n^3\pi^3$ for n odd and $b_n = 0$ for n even.

13. (a) $$u(x,t) = \frac{400}{\pi} \sum_{n \text{ odd}} \frac{1}{n} \exp\left(\frac{-n^2\pi^2 kt}{1600}\right)\sin \frac{n\pi x}{40}$$

    (b) With $k = 1.15$, $u(20,300) \approx 15.16°C$

    (c) About 19 hr 16 min

14. (a) The boundary value problem is

    $$u_t = ku_{xx} \quad (0 < x < 50)$$
    $$u_x(0,t) = u_x(50,t) = 0$$

$$u(x,0) = 2x$$

with $k = 1.15$ cm^2/sec for copper. By Equation (40) the solution is of the form

$$u(x,t) = a_0/2 + \sum a_n \exp(-n^2\pi^2 kt/2500)\cos n\pi x/50.$$

Consulting the Fourier series given in Equation (15) of Section 8.2, we satisfy the initial condition $u(x,0) = 2x$ by choosing $a_0 = 100$, $a_n = -400/n^2\pi^2$ for n odd, and $a_n = 0$ for n even.

(b) With the aid of the program

```
100 INPUT "X";X
110 INPUT "T";T
120 SUM = 50 : N = -1 : PI = 3.141593
130 A = 400/(PI*PI) : K = 1.15
140 B = PI*PI*K/2500
150 N = N + 2
160 TERM = A*EXP(-B*N*N*T)*COS(N*PI*X/50)/(N*N)
170 SUM = SUM - TERM
180 IF ABS(TERM) >.001 THEN GOTO 150
190 PRINT SUM
200 END
```

we find that $u(10,60) \approx 25.15$ (degrees Celsius)

(c) The program above makes a "hit or miss" approach effective. We find that

$u(10,300) = 41.60$ $\qquad$ $u(10,414.0) = 44.99$

$u(10,400) = 44.67$ $\qquad$ $u(10,415.0) = 45.02$

$u(10,410) = 44.90$

Thus $u = 45$ at $x = 10$ after 6 min 54.3 sec.

16. (a) Summing numerically the series in Problem 15 with $k = 0.15$ for iron, we find that $u(25,1800) \approx 21.92 \approx 22$ (degrees Celsius).

(b) Because for x fixed the temperature is a function of

the *product* kt, in the case of concrete slabs with k = 0.005 the same temperature will be attained when

$$(0.005)(t) = (0.15)(1800),$$

that is, when t = 54000 sec = 15 hr.

VIBRATING STRINGS AND
THE ONE-DIMENSIONAL WAVE EQUATION

In Problems 1-10 we use the general solution

$$y(x,t) = \sum (A_n \cos n\pi at/L + B_n \sin n\pi at/L) \sin n\pi x/L$$

of the string equation $y_{tt} = a^2 y_{xx}$ with endpoint conditions $y(0,t) = y(L,t) = 0$. This form of solution is obtained by superposition of the solutions in Equations (23) and (33) of Problems A and B in this section. It remains only to choose the coefficients $\{A_n\}$ and $\{B_n\}$ so as to satisfy given initial conditions $u(x,0) = f(x)$ and $u_t(x,0) = g(x)$.

1. Here $a = 2$ and $L = \pi$. To satisfy the condition $y(x,0) = (1/10)\sin 2x$ we choose $A_2 = 1/10$ and $A_n = 0$ otherwise. To satisfy the condition $y_t(x,0) = 0$ we choose $B_n = 0$ for all n. Thus

    $$y(x,t) = (1/10)\cos 4t \sin 2x.$$

2. Here $a = L = 1$. To satisfy the condition

    $$y(x,0) = (1/10)\sin \pi x - (1/20)\sin 3\pi x$$

we choose $A_1 = 1/10$, $A_3 = -1/20$, and $A_n = 0$ otherwise. To satisfy the condition $y_t(x,0) = 0$ we choose $B_n = 0$ for all n. Thus

$$y(x,t) = (1/10)\cos \pi t \sin \pi x - (1/20)\cos 3\pi t \sin 3\pi x.$$

3. Here $a = 1/2$ and $L = \pi$. Choosing $A_1 = 1/10$ and $A_n = 0$ otherwise, $B_1 = 1/5$ and $B_n = 0$ otherwise, we get

$$y(x,t) = (1/10)[\cos t/2 + 2 \sin t/2]\sin x.$$

4. Here $a = 1/2$ and $L = 2$, so $n\pi x/L = n\pi x/2$ and $n\pi at/L = n\pi t/4$. To satisfy the condition

$$y(x,0) = (1/5) \sin \pi x \cos \pi x$$
$$= (1/10)\sin 2\pi x = (1/10)\sin 4\pi x/2,$$

we choose $A_4 = 1/10$ and $A_n = 0$ for $n \neq 4$. To satisfy the condition $y_t(x,0) = 0$ we choose $B_n = 0$ for all n. Thus

$$y(x,t) = (1/10)\cos \pi t \sin 2\pi x.$$

5. Here $a = 5$ and $L = 3$. Choosing $A_3 = 1/4$ and $A_n = 0$ for $n \neq 3$, $B_6 = 1/\pi$ and $B_n = 0$ for $n \neq 6$, we get

$$y(x,t) = (1/4)\cos 5\pi t \sin \pi x + (1/\pi)\sin 10\pi t \sin 2\pi x.$$

6. Here $a = 10$ and $L = \pi$. To satisfy the condition $y_t(x,0) = 0$ we choose $B_n = 0$ for all n, so

$$y(x,t) = \sum A_n\cos 10nt \sin nx.$$

To satisfy the condition $y(x,0) = x(\pi - x)$ we choose $A_n = 8/\pi n^3$ for n odd and $A_n = 0$ for n even.

7. Here $a = 10$ and $L = 1$. To satisfy the condition $y(x,0) = 0$ we choose $A_n = 0$ for all n, so

$$y(x,t) = \sum B_n \sin 10 n\pi t \sin n\pi x.$$

To satisfy the condition $y_t(x,0) = x$ we choose

$$B_n = (-1)^{n+1}/(5\pi^2 n^2)$$

for $n \geq 1$ (see Equation (16) in Section 8.3).

8. Here $a = 2$ and $L = \pi$. To satisfy the condition $y(x,0) = \sin x$ we choose $A_1 = 1$ and $A_n = 0$ for $n > 1$, so

$$y(x,t) = \cos 2t \sin x + \sum B_n \sin 2nt \sin nx,$$

$$y_t(x,t) = -2 \sin 2t \sin x + \sum 2nB_n \cos 2nt \sin nx.$$

The condition $y_t(x,0) = 1$ will be satisfied if $2nB_n = 4/\pi n$ for n odd and $b_n = 0$ for n even. We therefore choose $B_n = 2/\pi n^2$ for n odd and $B_n = 0$ for n even.

9. Here $a = 2$ and $L = 1$. To satisfy the condition $y(x,0) = 0$ we choose $A_n = 0$ for all n, so

$$y(x,t) = \sum B_n \sin 2n\pi t \sin n\pi x.$$

To satisfy the condition $y_t(x,0) = x(1 - x)$ we choose $B_n = 4/n^4\pi^4$ for n odd, and $B_n = 0$ for n even.

10. Here $a = 5$ and $L = \pi$ so

$$y(x,t) = \sum (A_n \cos 5nt + B_n \sin 5nt) \sin nx.$$

We first compute the Fourier sine series

$$\sin^2 x = \sum b_n \sin nx$$

and find that $b_n = 0$ if n is even and

$$b_n = 2(n^2 + 4)/\pi n(4 - n^2)$$

if n is odd. To satisfy the condition $y(x,t) = \sin^2 x$ we choose $A_n = b_n$, and to satisfy the condition $y_t(x,t) = \sin^2 x$ we choose $B_n = b_n/5n$.

11. Substitution of L = 2 ft, T = 32 lb, and

$$\rho = (1/64) \text{ oz/ft} = 1/(32 \cdot 16 \cdot 64) \text{ slug/ft}$$

in Equation (26) in the text yields the fundamental frequency of 256 Hz. The velocity with which waves of this frequency move along the string is a = 1024 ft/sec.

18. When we separate variables as in Equations (8)-(12) in this section, we find that X(x) must satisfy the eigenvalue problem
$$X'' + \lambda X = 0, \qquad X(0) = X'(L) = 0.$$

In Example 4 of Section 2.10 we found that the eigenvalues and eigenfunctions of this problem are

$$\lambda_n = (2n - 1)^2 \pi^2 / 4L^2, \qquad X_n(x) = \sin(2n - 1)\pi x/2L$$

for n = 1,2,3$\cdots$. The function $T_n(t)$ must satisfy the conditions

$$T_n'' + \lambda_n a^2 T_n = 0, \qquad T_n'(0) = 0$$

so it follows that

$$T_n(t) = \cos(2n - 1)\pi at/2L.$$

Thus the form of y(t) is

$$y(x,t) = \sum A_n \cos(2n - 1)\pi at/2L \sin(2n - 1)\pi x/2L.$$

Finally, in order to satisfy the initial condition $y(x,0) = f(x)$ we use the odd half-multiple sine series

$$f(x) = \sum A_n \sin(2n - 1)\pi x/2L$$

discussed in Problem 21 of Section 8.3.

22. The eigenvalue problem

$$X'' + \lambda X = 0, \qquad X(0) = X(L) = 0$$

has the usual eigenvalues and eigenfunctions

$$\lambda_n = n^2\pi^2/L^2, \qquad X_n(x) = \sin n\pi x/L$$

for $n = 1,2,3,\cdots$. The function $T_n(t)$ satisfies the equation

$$T_n'' + (\omega_n)^2 T_n = 0, \qquad (\omega_n)^2 = (n^2\pi^2a^2/L^2) - h^2 > 0,$$

so

$$T_n(t) = A_n\cos \omega_n t + B_n\sin \omega_n t.$$

Thus

$$v(x,t) = \sum (A_n\cos \omega_n t + B_n\sin \omega_n t)\sin n\pi x/L.$$

To satisfy the conditions $v(x,0) = f(x)$ and $v_t(x,0) = h f(x)$ we choose $A_n = b_n$ and $B_n = hb_n/\omega_n$ where

$$f(x) = \sum b_n\sin n\pi x/L.$$

Then

$$v(x,t) = \sum (b_n/\omega_n)(\omega_n\cos \omega_n t + h \sin \omega_n t) \sin n\pi x/L$$

$$= \sum c_n\cos(\omega_n t - \alpha_n)\sin n\pi x/L$$

where

$$c_n = b_n/(\cos \alpha_n) \quad \text{and} \quad \alpha_n = \tan^{-1}(h/\omega_n).$$

Finally

$$y(x,t) = e^{-ht}v(x,t)$$

$$= e^{-ht} \sum c_n\cos(\omega_n t - \alpha_n)\sin n\pi x/L.$$

23. If $\pi/4 \leq x \leq 3\pi/4$ then

$\pi/2 \leq x + \pi/4 \leq \pi$ and $0 \leq x - \pi/4 \leq \pi/2$

so

$$y(x,\pi/4) = (1/2)[F(x + \pi/4) + F(x - \pi/4)]$$
$$= (1/2)[1 - \cos 2(x + \pi/4) + 1 - \cos 2(x - \pi/4)]$$
$$= (1/2)[1 - \cos (2x + \pi/2) + 1 - \cos (2x - \pi/2)]$$
$$= (1/2)[2 + \sin 2x - \sin 2x]$$
$$y(x,\pi/4) = 1$$

24. (a) $f''(x) = 4 \cos 2x = 0$ if $x = \pi/4$ or $x = 3\pi/4$.

(b) If $0 \leq t \leq \pi/4$ then $0 \leq \pi/4 \pm t \leq \pi/2$ so

$$y(\pi/4,t) = (1/2)[F(\pi/4 + t) + F(\pi/4 - t)]$$
$$= (1/2)[1 - \cos 2(\pi/4 + t) + 1 - \cos 2(\pi/4 - t)]$$
$$= (1/2)[1 - \cos (\pi/2 + 2t) + 1 - \cos (\pi/2 - 2t)]$$
$$= (1/2)[2 + \sin 2t - \sin 2t]$$
$$y(\pi/4,t) = 1$$

SECTION 8.7

STEADY-STATE TEMPERATURE AND LAPLACE'S EQUATION

1. Because $Y(0) = Y(b) = 0$ we take our separation of variables in the form

$$X'' - \lambda X = 0 = Y'' + \lambda Y$$

with $\lambda > 0$. Then it follows that

$$Y_n(y) = \sin n\pi y/b, \qquad \lambda_n = n^2\pi^2/b^2$$

and thence that

$$X_n(x) = A_n\cosh n\pi x/b + B_n\sinh n\pi x/b.$$

The condition that $X(0) = 0$ implies that $A_n = 0$ so $X_n(x) = C_n \sinh n\pi x/b$, and hence

$$u(x,y) = \sum C_n \sinh(n\pi x/b) \sin n\pi y/b.$$

Finally we satisfy the condition $u(a,y) = g(y)$ by choosing $C_n = b_n/(\sinh n\pi a/b)$ where $\{b_n\}$ are the Fourier sine coefficients of $g(y)$ on $0 \le y \le b$.

2. Because $Y(0) = Y(b) = 0$ we take our separation of variables in the form

$$X'' - \lambda X = 0 = Y'' + \lambda Y$$

with $\lambda > 0$. Then it follows that

$$Y_n(y) = \sin n\pi y/b, \qquad \lambda_n = n^2\pi^2/b^2$$

and thence that

$$X_n(x) = A_n \cosh n\pi x/b + B_n \sinh n\pi x/b.$$

The condition $X(a) = 0$ implies that

$$B_n = -\,(A_n \cosh n\pi a/b)/(\sinh n\pi a/b).$$

It now follows as in Equation (12) that

$$X_n(x) = C_n \sinh n\pi(a - x)/b,$$

so

$$u(x,t) = \sum C_n\, [\sinh n\pi(a - x)/b]\, \sin n\pi y/b.$$

Finally we satisfy the condition $u(0,y) = g(y)$ by choosing $C_n = b_n/(\sinh n\pi a/b)$ where $\{b_n\}$ are the Fourier sine coefficients of $g(y)$ on $0 \le y \le b$.

3. Just as in Example 1 of Section 8.7 we have $X_n(x) = \sin n\pi x/a$ and

$$Y_n(y) = A_n\cosh n\pi y/a + B_n\sinh n\pi y/a.$$

The condition $Y(0) = 0$ now yields $A_n = 0$ so $Y_n(y) = B_n\sinh n\pi y/a$, and hence

$$u(x,t) = \sum C_n\sin(n\pi x/a) \sinh n\pi y/a.$$

Finally we satisfy the condition $u(x,b) = f(x)$ by choosing $C_n = b_n/(\sinh n\pi b/a)$ where $\{b_n\}$ are the Fourier sine coefficients of $f(x)$ on $0 \le x \le a$.

4. We work with the separation of variables

$$X'' + \lambda X = 0 = Y'' - \lambda Y.$$

The eigenvalue problem

$$X'' + \lambda X = 0, \qquad\qquad X'(0) = X'(a) = 0$$

has eigenvalues and eigenfunctions $\lambda_0 = 0$, $X_0(x) = 1$ and

$$\lambda_n = n^2\pi^2/a^2, \qquad\qquad X_n(x) = \cos n\pi x/a$$

for $n = 1,2,3,\cdots$. When $n = 0$, $Y_0'' = 0$ yields $Y_0(y) = Ay + B$. Then $Y_0(0) = 0$ gives $B = 0$, so we take $Y_0(y) = y$. For $n > 0$ we have

$$Y_n(y) = A_n\cosh n\pi y/a + B_n\sinh n\pi y/a,$$

and $Y_n(0) = 0$ gives $A_n = 0$, so

$$u(x,y) = B_0 y + \sum B_n\cos n\pi x/a \sinh n\pi y/a.$$

Finally

$$u(y,b) = B_0 b + \sum (B_n\sinh n\pi b/a)\cos n\pi x/a,$$

so we satisfy the condition $u(x,b) = f(x)$ by taking $B_0 =$

$a_0/2b$ and $B_n = a_n/(\sinh n\pi b/a)$ where

$$f(x) = a_0/2 + \sum a_n\cos n\pi x/a.$$

5.   Now we work with the separation of variables

$$X'' - \lambda X = 0 = Y'' + \lambda Y.$$

The eigenvalue problem

$$Y'' + \lambda Y = 0, \qquad\qquad Y'(0) = Y'(b) = 0$$

has eigenvalues and eigenfunctions $\lambda_0 = 0$, $Y_0(y) = 1$ and

$$\lambda_n = n^2\pi^2/b^2, \qquad\qquad Y_n(y) = \cos n\pi y/b$$

for $n = 1,2,3,\cdots$.   When $n = 0$, $X_0'' = 0$ yields $X_0(x) = Ax + B$.   Then $X_0(a) = 0$ is satisfied by $X_0(x) = a - x$. For $n > 0$ we have

$$X_n(x) = A_n\cosh n\pi x/b + B_n\sinh n\pi x/b,$$

and $X_n(a) = 0$ is satisfied by the linear combination

$$X_n(x) = C_n\sinh n\pi(a - x)/b$$

of $\cosh n\pi x/b$ and $\sinh n\pi x/b$.   Therefore

$$u(x,t) = C_0(a - x) + \sum C_n\sinh n\pi(a - x)/b \cos n\pi y/b.$$

Finally we satisy the condition $u(0,y) = g(y)$ by choosing

$$C_0 = a_0/2a \quad\text{and}\quad C_n = b_n/(\sinh n\pi a/b)$$

where $\{a_n\}$ are the Fourier cosine coefficients of $g(y)$ on $0 \le y \le b$.

6. This is the same as Problem 4 except that $Y'(0) = 0$
    instead of $Y(0) = 0$, so $Y_0(y) = 1$ and $Y_n(y) =$
    $A_n \cosh n\pi y/a$ for $n > 0$. Then

$$u(x,y) = A_0 + \sum A_n \cos n\pi x/a \, \cosh n\pi y/a,$$

so we satisfy the condition $u(x,b) = f(x)$ by choosing A_0
$= a_0/2$ and $A_n = a_n/(\cosh n\pi b/a)$ where $\{a_n\}$ are the
Fourier cosine coefficients of $f(x)$ on $(0,a)$.

7. The eigenvalue problem

$$X'' + \lambda X = 0, \qquad X(0) = X(a) = 0$$

yields the eigenvalues and eigenfunctions

$$\lambda_n = n^2\pi^2/a^2, \qquad X_n(x) = \sin n\pi x/a$$

for $n = 1,2,3,\cdots$. Then

$$Y_n'' + \lambda_n Y_n = 0$$

yields

$$Y_n(y) = A_n e^{n\pi y/a} + B_n e^{-n\pi y/a}.$$

In order that $Y(y) \longrightarrow 0$ as $y \longrightarrow \infty$ we take $A_n = 0$, so

$$u(x,y) = \sum B_n e^{-n\pi y/a} \sin n\pi x/a.$$

Finally we satisfy the condition $u(x,0) = f(x)$ by choosing
the constants $\{B_n\}$ as the Fourier sine coefficients of
$f(x)$ on $0 \le x \le a$.

8. The eigenvalue problem

$$X'' + \lambda X = 0, \qquad\qquad X'0 = X'(a) = 0$$

yields $\lambda_0 = 0$, $X_0(x) = 1$ and

$$\lambda_n = n^2\pi^2/a^2, \qquad\qquad X_n(x) = \cos n\pi x/a$$

for $n > 0$. Then

$$Y_n" + \lambda_n Y_n = 0$$

yields $Y_0(y) = Ay + B$ and

$$Y_n(y) = A_n e^{-n\pi y/a} + B_n e^{n\pi y/a}.$$

In order that $Y(y)$ be bounded as $y \longrightarrow \infty$, we take $A_n = 0$ and $B_n = 0$ for $n > 0$, so

$$u(x,y) = B + \sum A_n e^{-n\pi y/a} \cos n\pi x/a.$$

Finally we satisfy the condition $u(x,0) = f(x)$ by choosing $B = a_0/2$ and $A_n = a_n$ where $\{a_n\}$ are the Fourier cosine coefficients of $f(x)$ on $(0,a)$.

9. $u(0,5) \approx 41.53$; $u(5,5) = 50$; $u(10,5) \approx 58.47$

10. The boundary value problem is

$$u_{xx} + u_{yy} = 0 \quad (0 < x < a, \quad 0 < y < b)$$
$$u(0,y) = u_x(a,y) = u(x,0) = 0,$$
$$u(x,b) = f(x).$$

The eigenvalue problem

$$X" + \lambda X = 0, \qquad\qquad X(0) = X'(a) = 0$$

yields (by Example 4 in Section 2.10)

$$\lambda_n = (2n - 1)^2\pi^2/4a^2, \qquad X_n(x) = \sin(2n - 1)\pi x/2a$$

for $n = 1,2,3,\cdots$. Then

$$Y_n'' - \lambda_n Y_n = 0$$

yields

$$Y_n(y) = A_n \cosh(2n - 1)\pi y/2a + B_n \sinh(2n - 1)\pi y/2a.$$

Because $Y(0) = 0$, we choose $A_n = 0$, so

$$u(x,y) = \sum B_n \sin (2n - 1)\pi x/2a \sinh (2n - 1)\pi y/2a.$$

Finally we satisfy the condition $u(x,b) = f(x)$ by choosing $B_n = b_{2n-1}/\sinh(2n -1)\pi b/2a$, where $\{b_{2n-1}\}$ are the odd half-multiple sine coefficients of $f(x)$ on $(0,a)$, as given by Problem 21 in Section 8.3.

11. $u(x,y) = \sum C_n \sinh n\pi(a - x)/2b \cos n\pi y/2b$

where $C_n = 0$ for n even, while if n is odd then $C_n = a_n/(\sinh n\pi a/2b)$ where

$$a_n = \frac{2}{b} \int_0^b g(y) \cos \frac{n\pi y}{2b} \, dy.$$

12. The boundary value problem is

$$u_{xx} + u_{yy} = 0 \quad (0 < x < 30, \quad y > 0)$$
$$u(0,y) = u_x(30,y) = 0$$
$$u(x,y) \quad \text{bounded as} \quad y \longrightarrow \infty$$
$$u(x,0) = 25$$

The eigenvalue problem

$$X'' + \lambda X = 0, \qquad X(0) = X'(30) = 0$$

yields (by Example 4 in Section 2.10)

$$\lambda_n = (2n - 1)^2 \pi^2/3600, \qquad X_n(x) = \sin(2n -1)\pi x/60$$

for $n = 1,2,3,\cdots$. Then

$$Y_n" - \lambda_n Y_n = 0$$

yields

$$Y_n(y) = A_n e^{n\pi y/60} + B_n e^{-n\pi y/60}$$

and we take $A_n = 0$ in order that $Y_n(y)$ be bounded as $y \longrightarrow \infty$. Hence

$$u(x,y) = \sum B_n e^{-n\pi y/60} \sin(2n - 1)\pi x/60.$$

In order to obtain the formula for $u(x,y)$ given in the text it remains only to compute (Section 8.3, Problem 21) the odd half-multiple Fourier sine coefficients of $u(x,0) = 25$ on $[0,30]$.

13. $C_n = b_n/a^n$ where $\{b_n\}$ are the Fourier sine coefficients off(θ) on $0 \le \theta \le \pi$.

14. When we introduce the separation of variables $u(r,\theta) = R(r)T(\theta)$, we get

$$T_0(\theta) = 1, \qquad R_0(r) = C_0 + D_0 \ln r,$$
$$T_n(\theta) = A_n \cos n\theta + B_n \sin n\theta,$$
$$R_n(r) = C_n r^n + D_n r^{-n}$$

just as in Equations (25)-(30) in the text. We choose $B_n = 0$ in order that $T'(0) = T'(\pi) = 0$, and $D_n = 0$ in order that $R(r)$ be continuous at $r = 0$. Hence

$$u(r,\theta) = c_0/2 + \sum c_n r^n \cos n\theta,$$

and we satisfy the condition $u(a,\theta) = f(\theta)$ by choosing $c_0 = a_0$, $c_n = a_n/a^n$ where $\{a_n\}$ are the Fourier cosine coefficients of $f(\theta)$ on $(0,\pi)$.

15. $C_n = \dfrac{2}{\pi a^{n/2}} \displaystyle\int_0^\pi f(\theta) \sin \dfrac{n\theta}{2} \, d\theta$

16. The only difference between the exterior problem here and
     the interior problem in the text is that in

$$R_n(r) = C_n r^n + D_n r^{-n}$$

we must choose $C_n = 0$ in order that $R_n(r)$ be bounded as
$r \longrightarrow \infty$.

20. When we substitute $v(r,t) = r\, u(r,t)$ we get the boundary
     value problem

$$v_t = kv_{rr} \qquad (r < a, \quad t > 0)$$
$$v(0,t) = v(a,t) = 0$$
$$v(r,0) = T_0 r$$

that corresponds to a heated rod along the interval $0 \le r \le$
a.   It therefore follows from Equation (31) in Section 7.5
that

$$v(r,t) = \sum b_n \exp(-n^2\pi^2 kt/a^2)\sin n\pi x/a.$$

To get the formula given in the text it remains only to
calculate the Fourier sine coefficients $\{b_n\}$ of $f(r) =$
$T_0 r$ on $0 < r < a$,   and finally to divide $v(r,t)$ by r to
get $u(r,t)$.

21. (a) Since we cannot simply substitute $r = 0$,   we apply
     continuity of $u(r,t)$ at $r = 0$ and calculate

$$u(0,t) = \lim_{r \longrightarrow 0} u(r,t),$$

noting that

$$\lim_{r \longrightarrow 0} \frac{\sin n\pi r/a}{r} = \frac{nr}{a}.$$

EIGENVALUES AND BOUNDARY VALUE PROBLEMS

SECTION 9.1

STURM-LIOUVILLE PROBLEMS AND EIGENFUNCTION EXPANSIONS

1.   In the notation of Equation (9) in this section we have
$\alpha_1 = \beta_1 = 0$ and $\alpha_2 = \beta_2 = 1$,   so Theorem 1 implies that the
eigenvalues are all nonnegative.   If $\lambda = 0$,   then
$y" = 0$ implies that $y(x) = Ax + B$.   Then $y'(x) = A$,   so
the endpoint conditions yield $A = 0$,   but B remains
arbitrary.   Hence $\lambda_0 = 0$ is an eigenvalue with
eigenfunction

$$y_0(x) = 1.$$

If $\lambda = \alpha^2 > 0$,   then the equation $y" + \alpha^2 y = 0$ has
general solution

$$y(x) = A \cos \alpha x + B \sin \alpha x,$$
with
$$y'(x) = -A\alpha \sin \alpha x + B\alpha \cos \alpha x.$$

Then $y'(0) = 0$ yields $B = 0$ so $A \neq 0$,   and then

$$y'(L) = -A\alpha \sin \alpha L = 0,$$

so αL must be an integral multiple of π.   Thus the   nth
positive eigenvalue is

$$\lambda_n = \alpha_n^2 = n^2\pi^2/L^2,$$

and the associated eigenfunction is

$$y_n(x) = \cos n\pi x/L.$$

2. In the notation of Equation (9) in this section we have
    $\alpha_1 = \beta_2 = 1$ and $\alpha_2 = \beta_1 = 0$, so Theorem 1 implies that the
    eigenvalues are all nonnegative. If $\lambda = 0$, then
    $y'' = 0$ implies $y(x) = Ax + B$. But then $y(0) = B = 0$ and
    $y'(L) = A = 0$, so it follows that 0 is not an eigenvalue.
    We may therefore write $\lambda = \alpha^2 > 0$, so our equation is
    $y'' + \alpha^2 y = 0$ with general solution

$$y(x) = A \cos \alpha x + B \sin \alpha x.$$

Now $y(0) = A = 0$, so $y(x) = B \sin \alpha x$ and

$$y'(x) = B\alpha \cos \alpha x.$$

Hence
$$y'(L) = B\alpha \cos \alpha L = 0,$$

so it follows that αL must be an odd multiple of $\pi/2$.
Thus
$$\alpha_n = (2n - 1)\pi/2L, \quad \lambda_n = (\alpha_n)^2, \quad y_n(x) = \sin \alpha_n x.$$

3. If $\lambda = 0$ then $y'' = 0$ yields $y(x) = Ax + B$ as usual.
    But $y'(0) = A = 0$, and then $hy(L) + y'(L) = h(B) + 0 = 0$,
    so $B = 0$ also. Thus $\lambda = 0$ is not an eigenvalue. If
    $\lambda = \alpha^2 > 0$ so our equation is $y'' + \alpha^2 y = 0$, then

$$y(x) = A \cos \alpha x + B \sin \alpha x,$$

$$y'(x) = -A\alpha \sin \alpha x + B\alpha \cos \alpha x.$$

Now $y'(0) = 0$ yields $B = 0$, so we may write

$$y(x) = \cos \alpha x, \qquad y'(x) = -\alpha \sin \alpha x.$$

The equation

$$hy(L) + y'(L) = h \cos \alpha L - \alpha \sin \alpha L = 0$$

then gives

$$\tan \alpha L = h/\alpha = hL/\alpha L,$$

so $\beta_n = \alpha_n L$ is the nth positive root of the equation

$$\tan x = hL/x.$$

Thus

$$\lambda_n = \alpha_n^2 = \beta_n^2/L^2, \quad y_n(x) = \cos \beta_n x/L.$$

Finally, a sketch of the graphs $y = \tan x$ and $y = hL/x$ indicates that $\beta_n \approx (n - 1)\pi$ for n large.

4. Here $\alpha_1 = h > 0$, $\alpha_2 = \beta_1 = 1$, and $\beta_2 = 0$, so by Theorem 1 there are no negative eigenvalues. If $\lambda = 0$ and $y(x) = Ax + B$, then the equations

$$hy(0) - y'(0) = hB - A = 0, \qquad y(L) = AL + B = 0$$

imply $h = A/B = -1/L < 0$. Thus 0 is not an eigenvalue. If $\lambda = \alpha^2 > 0$ and

$$y(x) = A \cos \alpha x + B \sin \alpha x,$$

then the condition $h\,y(0) = y'(0)$ yields $B = hA/\alpha$, so

$$y(x) = (A/\alpha)(\alpha \cos \alpha x + h \sin \alpha x)$$
$$= (A/\beta)(\beta \cos \beta x/L + hL \sin \beta x/L)$$

where $\beta = \alpha L$. Then the condition

$$y(L) = (A/B)(\beta \cos \beta + hL \sin \beta) = 0$$

reduces to $\tan \beta = -\beta/hL$.

6.　$y_n(x) = \sin(2n - 1)\pi x/2L$　so Equation (25) yields

$$c_n \int_0^L \sin^2(2n - 1)\pi x/2L\ dx = \int_0^L f(x)\sin(2n - 1)\pi x/2L\ dx.$$

When we evaluate the integral on the left we immediately get the formula in (29).

7.　The coefficient c_n in (23) is given by Formula (25) with $f(x) = r(x) = 1$, $a = 0$, $b = L$, and $y_n(x) = \sin \beta_n x/L$. Using the fact that $\tan \beta_n = -\beta_n/hL$, so $(\sin \beta_n)/\beta_n = -(\cos \beta_n)/hL$, we find that

$$\int_0^L \sin^2 \beta_n x/L\ dx = (hL + \cos^2 \beta_n)/2h$$

and

$$\int_0^L \sin \beta_n x/L\ dx = L(1 - \cos \beta_n)/\beta_n.$$

Hence the desired eigenfunction expansion is

$$1 = 2hL \sum [(1 - \cos \beta_n)/\beta_n(hL + \cos^2 \beta_n)]\sin \beta_n x/L.$$

for　$0 < x < L$.

8.　The coefficient c_n in (23) is given by Formula (25) with $f(x) = r(x) = 1$, $a = 0$, $b = L$, and $y_n(x) = \cos \beta_n x/L$. The result is

$$c_n = (4 \sin \beta_n)/(2\beta_n + \sin 2\beta_n)$$

for　$n = 1,2,3,\cdots$, so the desired eigenfunction expansion is

$$1 = \sum (4 \sin \beta_n \cos \beta_n x/L)/(2\beta_n + \sin 2\beta_n).$$

9.　The coefficient c_n in (23) is given by Formula (25) with

$f(x) = r(x) = 1$, $\quad a = 0$, $\quad b = 1$, $\quad$ and $\quad y_n(x) = \sin \beta_n x$.
Using the fact that $\tan \beta_n = -\beta_n/h$, $\quad$ so $\quad h \sin \beta_n = -\beta_n \cos \beta_n$, $\quad$ we find that

$$\int_0^1 \sin^2 \beta_n x \, dx = (h + \cos^2 \beta_n)/2h$$

and

$$\int_0^1 x \sin \beta_n x \, dx = (\sin \beta_n - \beta_n \cos \beta_n)/\beta_n^2.$$

$$= (1 + h)(\sin \beta_n)/\beta_n^2.$$

It follows that the desired expansion is given by

$$x = 2h(1 + h) \sum [(\sin \beta_n)/\beta_n^2(h + \cos^2\beta_n)]\sin \beta_n x$$

for $\quad 0 < x < 1$.

10. The coefficient c_n in (23) is given by Formula (25) with $f(x) = x$, $\quad r(x) = 1$, $\quad a = 0$, $\quad b = 1$, $\quad$ and $\quad y_n(x) = \cos \beta_n x$. The result is

$$c_n = 4(\beta_n \sin \beta_n + \cos \beta_n - 1)/\beta_n(2\beta_n + \sin 2\beta_n).$$

With this value of c_n for $n = 1,2,3,\cdots$ the desired eigenfunction expansion is

$$x = \sum c_n \cos \beta_n x/L$$

11. If $\lambda = 0$ then $y'' = 0$ implies that $y(x) = Ax + B$. Then $y(0) = 0$ gives $B = 0$, so $y(x) = Ax$. Hence

$$hy(L) - y'(L) = h(AL) - A = A(hL - 1) = 0$$

if and only if $hL = 1$, in which case $\lambda_0 = 0$ has associated eigenfunction $y_0(x) = x$.

12. If $\lambda = -\alpha^2 < 0$, then the general solution of $y'' - \alpha^2 y =$

0   is

$$y(x) = A \cosh \alpha x + B \sinh \alpha x.$$

But then $y(0) = A = 0$, so we may take $y(x) = \sinh \alpha x$. Now the condition $hy(L) = y'(L)$ yields

$$h \sinh \alpha L = \alpha \cosh \alpha L.$$

It follows that $\beta = \alpha L$ must be a root of the equation

$$\tanh x = x/hL.$$

The curve $y = \tanh x$ passes through the origin with slope 1, and is concave upward for $x < 0$, concave downward for $x > 0$. Hence this curve and the line $y = x/hL$ intersect other than at the origin if and only if the slope of the line is less than 1, that is, if and only if $hL > 1$. In this case, with β_0 the positive root of $\tanh x = x/hL$, we have $\lambda_0 = -(\beta_0)^2$ and $y_0(x) = \sinh \beta_0 x$.

13.   The computation with $\lambda = +\alpha^2 > 0$ is essentially the same as in Problem 12, replacing hyperbolic functions with ordinary trigonometric functions.

14.   With $\lambda = 0$, $y'' = 0$, and hence $y(x) = Ax + B$, we have $y(0) = B = 0$, so $y(x) = Ax$. Then the condition $hy(L) = y'(L)$ reduces to the equation $hL = A$, which is satisfied because $hL = 1$. Thus $\lambda_0 = 0$ is an eigenvalue with associated eigenfunction $y_0(x) = x$. Together with the positive eigenvalues and associated eigenfunctions provided by Problem 13, this gives the eigenfunction expansion

$$f(x) = c_0 x + \sum c_n \sin \beta_n x/L$$

where $\tan \beta_n = \beta_n$. The given formulas for the coefficients follow readily upon evaluation of the denominator integrals in (25).

18. By Equation (16) in Section 8.3,

$$bx = (2bL/\pi) \sum [(-1)^{n+1}/n] \sin n\pi x/L.$$

If $y = \sum b_n \sin n\pi x/L$, then

$$EIy^{(4)} = EI \sum (n^4\pi^4 b_n/L^4) \sin n\pi x/L.$$

When we equate the coefficients of $\sin n\pi x/L$ in these two series, we get

$$b_n = 2bL^5(-1)^{n+1}/EIn^5\pi^5$$

as desired.

20. With $\lambda = \alpha^4$, the general solution of $y^{(4)} - \alpha^4 y = 0$ is

$$y(x) = A \cosh \alpha x + B \sinh \alpha x + C \cos \alpha x + D \sin \alpha x,$$

and then

$$y'(x) = \alpha(A \sinh \alpha x + B \cosh \alpha x - C \sin \alpha x + D \cos \alpha x).$$

The conditions $y(0) = 0$ and $y'(0) = 0$ yield $C = -A$ and $D = -B$, so now

$$y(x) = A(\cosh \alpha x - \cos \alpha x) + B(\sinh \alpha x - \sin \alpha x).$$

The conditions $y''(L) = 0$ and $y^{(3)}(L) = 0$ yield the two linear equations

$$A(\cosh \alpha L + \cos \alpha L) + B(\sinh \alpha L + \sin \alpha L) = 0,$$
$$A(\sinh \alpha L - \sin \alpha L) + B(\cosh \alpha L + \cos \alpha L) = 0.$$

This linear system can have a non-trivial solution for A and B only if its coefficient determinant vanishes,

$$(\cosh \alpha L + \cos \alpha L)^2 - (\sinh^2\alpha L - \sin^2\alpha L) = 0.$$

This equation simplifies to

$$\cosh \alpha L \cos \alpha L + 1 = 0,$$

so $\beta = \alpha L$ satisfies the equation

$$\cosh x \cos x = -1.$$

The eigenvalue corresponding to the nth root β_n is

$$\lambda_n = (\alpha_n)^4 = (\beta_n/L)^4.$$

Finally the first equation in the pair above yields

$$B = -A(\cosh \alpha L + \cos \alpha L)/(\sinh \alpha L + \sin \alpha L),$$

so we may take

$$y_n(x) = (\sinh \beta_n + \sin \beta_n)(\cosh \beta_n x/L - \cos \beta_n x/L)$$
$$- (\cosh \beta_n + \cos \beta_n)(\sinh \beta_n x/L - \sin \beta_n x/L)$$

as the eigenfunction associated with λ_n.

SECTION 9.2

APPLICATIONS OF EIGENFUNCTION SERIES

1. Most of the work required here has already been done in Problem 3 of Section 9.1, where we saw that the Sturm-Liouville problem

$$X'' + \alpha^2 X = 0, \qquad X'(0) = hX(L) + X'(L) = 0$$

has eigenfunctions

$$X_n(x) = \cos \beta_n x/L$$

for $n = 1, 2, 3, \cdots,$ with $\{\beta_n\}$ being the positive roots of the equation $\tan x = hL/x$. For the coefficients in the eigenfunction expansion we need only calculate the integral

$$\int_0^L \cos^2 \beta_n x/L \, dx = (hL + \sin^2 \beta_n)/2h.$$

2. $u(x,y) = \sum c_n \sinh \beta_n (L - y)/L \sin \beta_n x/L$

where $\{\beta_n\}$ are the positive roots of the equation $\tan x = - x/hL$ and

$$c_n = \frac{4\beta_n}{L(\sinh \beta_n)(2\beta_n - \sin 2\beta_n)} \int_0^L f(x) \sin \frac{\beta_n x}{L} \, dx$$

3. $u(x,y) = \sum c_n \sinh \beta_n (L - x)/L \cos \beta_n y/L$

where $\{\beta_n\}$ are the positive roots of the equation $\tan x = x/hL$ and

$$c_n = \frac{2h}{(\sinh \beta_n)(hL + \sin^2 \beta_n)} \int_0^L g(y) \cos \frac{\beta_n y}{L} \, dy$$

4. $u(x,y) = \sum c_n \exp(-\beta_n y/L) \sin \beta_n x/L$

where $\{\beta_n\}$ are the positive roots of the equation $\tan x = - x/hL$ and

$$c_n = \frac{4\beta_n}{L(2\beta_n - \sin 2\beta_n)} \int_0^L f(x) \sin \frac{\beta_n x}{L} \, dx$$

5. $u(x,t) = \sum c_n \exp(-\beta_n kt/L^2) X_n(x)$

where $\{\beta_n\}$ are the positive roots of the equation $\tan x = - x/hL,$

$$X_n(x) = \beta_n \cos \beta_n x/L + hL \sin \beta_n x/L,$$

and

$$c_n \int_0^L [X_n(x)]^2 \, dx = \int_0^L f(x) X_n(x) \, dx$$

6. $u(x,t) = \sum c_n [\exp(-(\beta_n)^2 kt/L^2)][\beta_n \cos \beta_n x/L + hL \sin \beta_n x/L]$

where $\{\beta_n\}$ are the roots of the equation

$$\tan x = 2hLx/(x^2 - h^2 L^2).$$

7. Four terms of the series for $u(x,y)$ give $u(1,1) \approx 31.38°C$.

8. Starting with the general solution

$$y = A \cos \alpha x + B \sin \alpha x$$

of $y'' + \alpha^2 y = 0$, the condition $y'(0) = 0$ gives $B = 0$, so

$$y = A \cos \alpha x, \qquad y' = -\alpha A \sin \alpha x.$$

Then the condition $My(L) + mLy'(L) = 0$ gives

$$MA \cos \alpha L - mL\alpha A \sin \alpha L = 0,$$

which is equivalent to the equation $\tan x = M/mx$ with $x = \alpha L$.

10. (a) With $\delta = 7.75$ gm/cm^3 and $E = 2 \cdot 10^{12}$ in Equation (16), the speed of sound in steel is

$$a = \sqrt{(E/\delta)} \approx 5.08 \times 10^5 \text{ cm/sec} \approx 11364 \text{ mph}.$$

(b) With $\delta = 1$ gm/cm^3 and $K = 2.25 \cdot 10^{10}$ in Equation (16), the speed of sound in water is

$$a = \sqrt{(K/\delta)} \approx 1.50 \times 10^5 \text{ cm/sec} \approx 3355 \text{ mph}.$$

12. The boundary value problem is

$$u_{tt} = a^2 u_{xx} \quad (0 < x < L, \ t > 0)$$
$$u(0,t) = ku(L,t) + AEu_x(L,t) = 0$$
$$u(x,0) = f(x), \quad u_t(x,0) = 0$$

Starting with the general solution

$$X(x) = A \cos \alpha x + B \sin \alpha x$$

of $X'' + \alpha^2 X = 0$, the condition $X(0) = 0$ gives $A = 0$, so

$$X(x) = \sin \alpha x, \qquad X'(x) = \alpha \cos \alpha x.$$

Then the condition $kX(L) + AEX'(L) = 0$ yields

$$k \sin \alpha L + AE\alpha \cos \alpha L = 0,$$

which is equivalent to the equation

$$\tan x = - AEx/kL$$

with $x = \alpha L$, $\alpha = x/L$. If $\{\beta_n\}$ are the positive roots of this equation, then the nth eigenvalue is $\lambda_n = (\alpha_n)^2 = (\beta_n/L)^2$ with associated eigenfunction

$$X_n(x) = \sin \beta_n x/L.$$

The associated function of t is

$$T_n(t) = A_n \cos \beta_n at/L + B_n \sin \beta_n at/L,$$

but the condition $T'(0) = 0$ yields $B_n = 0$. Hence we obtain a solution of the form

$$u(x,t) = \sum c_n \cos \beta_n at/L \sin \beta_n x/L.$$

16. When we substitute $v(r,t) = ru_r(r,t)$ we get the boundary value problem

$$v_t = kv_{rr}$$
$$v(0,t) = v(a,t) - av_r(a,t) = 0$$
$$v(r,0) = r\,f(r)$$

then $v(r,t) = R(r)T(t)$ yields the equations

$$R'' + \lambda R = 0, \qquad T' = -\lambda kT.$$

If $\lambda_0 = 0$ then $R(r) = Ar + B$. The condition $R(0) = 0$ gives $B = 0$, and $R(r) = Ar$ satisfies the condition $R(a) - aR'(a) = 0$. Thus $\lambda_0 = 0$ is an eigenvalue with eigenfunction

$$R_0(r) = r; \qquad T_0(t) = 1.$$

If $\lambda = \alpha^2 > 0$ then

$$R(r) = A \cos \alpha r + B \sin \alpha r$$

and $R(0) = 0$ gives $A = 0$, so

$$R(r) = \sin \alpha r, \qquad R'(r) = \alpha \cos \alpha r.$$

The condition $R(a) = a\,R'(a)$ yields $\sin \alpha a = a\alpha \cos \alpha a$, that is,

$$\tan x = x$$

where $x = \alpha a$. If $\{\beta_n\}$ are the roots of this equation, then $\lambda_n = (\beta_n/a)^2$ is an eigenvalue with associated eigenfunction

$$R_n(r) = \sin \beta_n r/a; \qquad T_n(t) = \exp(-(\beta_n)^2 kt/a^2).$$

We therefore obtain a solution of the form

$$v(r,t) = c_0 r + \sum c_n \exp[-(\beta_n)^2 kt/a^2]\sin \beta_n r/L.$$

The formulas given for the coefficients follow immediately from Problem 14 in Section 9.1, and finally we obtain $u(r,t)$ upon division of $v(r,t)$ by r.

18. The only difference from Example 3 is that the solution of Equation (37) with $T_n'(0) = 0$ is $T_n(t) = \sin n^2\pi^2 a^2 t/L^2$.

20. The fundamental frequency is

$$\omega_1 = \pi^2 a^2/L^2 = (\pi^2/L^2)\sqrt{(EI/\rho)}.$$

with

$$E = 2 \cdot 10^{12} \ \text{dyne/cm}^2,$$
$$I = (2.54)^4/12 \approx 3.47 \ \text{cm}^4,$$
$$\rho = (7.75)(2.54)^2 \approx 50.00 \ \text{gm/cm},$$
$$L = (19)(2.54) \approx 48.26 \ \text{cm},$$

we calculate

$$\omega_1 \approx 1578 \ \text{rad/sec} \approx 251 \ \text{cycles/sec}.$$

Thus we hear middle C (approximately).

SECTION 9.3

STEADY PERIODIC SOLUTIONS AND NATURAL FREQUENCIES

In Problems 1-6 we substitute $u(x,t) = X(x)\cos \omega t$ in

$$u_{tt} = a^2 u_{xx} \qquad (a^2 = E/\delta)$$

and then cancel the factor $\cos \omega t$ to obtain the ordinary differential equation

$$a^2 X'' + \omega^2 X = 0$$

with general solution

$$X(x) = A \cos \omega x/a + B \sin \omega x/a.$$

It then remains only to apply the given endpoint conditions to determine the natural (circular) frequencies -- the values of ω for which a non-trivial solution exists.

1. With endpoint conditions $X(0) = X(L) = 0$ the nth natural frequency is $\omega_n = (n\pi/L)\sqrt{(E/\delta)}$.

2. Endpoint conditions: $X'(0) = X'(L) = 0$

    The condition $X'(0) = 0$ gives $B = 0$, so we have

    $$X(x) = \cos \omega x/a, \quad X'(x) = -(\omega/a)\sin \omega x/a.$$

    Hence the condition $X'(L) = 0$ implies that $\omega L/a$ is an integral multiple of π. Thus the nth natural frequency is

    $$\omega_n = n\pi a/L = (n\pi/L)\sqrt{(E/\delta)}.$$

3. With endpoint conditions $X(0) = X'(L) = 0$ the nth natural frequency is $\omega_n = [(2n - 1)\pi/2L]\sqrt{(E/\delta)}$.

4. Endpoint conditions: $u(0,t) = mu_{tt}(L,t) + AEu_x(L,t) = 0$

    The nth natural frequency is

    $$\omega_n = (\beta_n/L)\sqrt{(E/\delta)}$$

    where β_n is the nth positive root of the equation

    $$mx \tan x = M.$$

    This is the special case $k = 0$ of Problem 7 below.

5.	Endpoint conditions:	$u_x(0,t) = ku(L,t) + AEu_x(L,t) = 0$

The nth natural frequency is

$$\omega_n = (\beta_n/L)\sqrt{(E/\delta)}$$

where β_n is the nth positive root of the equation

$$AEx \tan x = kL.$$

6.	Endpoint conditions:

$$m_0 u_{tt}(0,t) - AEu_x(0,t) = 0,$$
$$m_1 u_{tt}(L,t) + AEu_x(L,t) = 0$$

When we substitute $u(x,t) = X(t)\cos \omega t$ in the two endpoint conditions and then cancel the $\cos \omega t$ factor, we get the equations

$$m_0 \omega^2 X(0) + KX'(0) = 0$$
$$m_1 \omega^2 X(L) - KX'(L) = 0$$

where we write $K = AE$ to avoid confusion with the coefficient of $\cos \omega x/a$ in

$$X(x) = A \cos \omega x/a + B \sin \omega x/a.$$
Then
$$X(0) = A, \qquad X'(0) = B\omega/a$$
$$X(L) = A \cos \omega L/a + B \sin \omega L/a$$
$$X'(L) = (\omega/a)(- A \sin \omega L/a + B \cos \omega L/a)$$

If we write $z = \omega L/a$, then

$$X(0) = A, \qquad X'(0) = Bz/L$$
$$X(L) = A \cos z + B \sin z$$
$$X'(L) = (z/L)(- A \sin z + B \cos z)$$

When we substitute these values and $\omega = az/L$ in the two

endpoint conditions above and collect coefficients of   A
and B,   we get the equations

$$m_0 a^2 z A + KLB = 0,$$

$$A(m_1 a^2 z \cos z + KL \sin z) + B(m_1 a^2 z \sin z - KL \cos z) = 0.$$

In order for this system to have a non-trivial solution for
A   and   B,   its determinant of coefficients must vanish,

$$m_0 a^2 z (m_1 a^2 z \sin z - KL \cos z)$$
$$- KL(m_1 a^2 z \cos z + KL \sin z) = 0.$$

When we substitute $a^2 = E/\delta$, $M = \delta AL$,   and $K = AE$,   this
last equation simplifies finally to the frequency equation

$$(m_0 m_1 z^2 - M^2) \sin z = M(m_0 + m_1) z \cos z.$$

If β_n is the nth positive root, then the nth natural
frequency is

$$\omega_n = (\beta_n / L) \sqrt{(E/\delta)}.$$

7.   Endpoint conditions:

$$u(0,t) = mu_{tt}(L,t) + AEu_x(L,t) + ku(L,t) = 0$$

The condition $u(0,t) = 0$ implies that

$$X(x) = \sin \omega x/a, \qquad X'(x) = (\omega/a) \cos \omega x/a.$$

When we substitute $u(x,t) = X(x) \cos \omega t$ in the endpoint
condition at $x = L$ and cancel the $\cos \omega t$ factor we get

$$- m\omega^2 X(L) + AEX'(L) + kX(L) = 0.$$

Next we substitute

$$z = \omega L/a, \qquad \omega = az/L, \qquad a^2 = E/\delta.$$

$$X(L) = \sin z, \qquad X'(L) = (z/L) \cos z.$$

The result simplifies readily to the frequency equation

$$(mEz^2 - k\delta L^2)\sin z = MEz \cos z.$$

If β_n is the nth positive root, then the nth natural frequency is $\omega_n = (\beta_n/L)\,\sqrt{(E/\delta)}$.

In Problems 8-14 we substitute $y(x,t) = X(x)\cos \omega t$ in

$$y_{tt} + a^4 y_{xxxx} = 0 \qquad (a^4 = EI/\rho)$$

and then cancel the factor $\cos \omega t$ to obtain the ordinary differential equation

$$a^4 X^{(4)} - \omega^2 X = 0$$

with general solution

$$X(x) = A \cosh \theta x/a + B \sinh \theta x/a + C \cos \theta x/a + D \sin \theta x/a.$$

where $\theta = \sqrt{\omega}$. We then get the natural frequencies of vibration by applying the given endpoint conditions.

8. Endpoint conditions: $y(0,t) = y_{xx}(0,t) = 0$
$$y(L,t) = y_{xx}(L,t) = 0$$

Just as in Example 3 of Section 9.2, the conditions $X(0) = X''(0) = 0$ imply that $A = C = 0$ so

$$X(L) = B \sinh \theta L/a + D \sin \theta L/a = 0,$$
$$X''(L) = (\theta^2/a^2)(B \sinh \theta L/a - D \sin \theta L/a) = 0.$$

It follows that

$$B \sinh \theta L/a = D \sin \theta L/a = 0.$$

But sinh $\theta L/a \neq 0$ so $B = 0$. Hence $D \neq 0$ so $\sin \theta L/a = 0$. Thus $\theta L/a = n\pi$, an integral multiple of π. Therefore the nth natural frequency $\omega = \theta^2$ is

$$\omega_n = n^2\pi^2 a^2/L^2 = (n^2\pi^2/L^2)\sqrt{(EI/\rho)}.$$

10. Endpoint conditions: $y(0,t) = y_x(0,t) = 0$
$$y_{xx}(L,t) = y_{xxx}(L,t) = 0$$

Here we have the equation

$$X^{(4)} - \lambda X = 0$$

with $\lambda = \omega^2/a^4$ and endpoint conditions

$$X(0) = X'(0) = X''(L) = X^{(3)}(L) = 0.$$

According to Problem 20 in Section 9.1 the nth eigenvalue is

$$\lambda_n = (\omega_n/a^2)^2 = (\beta_n/L)^4$$

where $\{\beta_n\}$ are the positive roots of

$$\cosh x \cos x = -1.$$

Thus the nth natural frequency is

$$\omega_n = (\beta_n/L)^2 a^2 = (\beta_n/L)^2\sqrt{(EI/\rho)}.$$

12. This problem is the special case $k = 0$ of Problem 14 which follows.

14. Endpoint conditions:

$$y(0,t) = y_x(0,t) = y_{xxx}(L,t) = 0$$
$$my_{tt}(L,t) = EIy_{xxx}(L,t) - ky(L,t)$$

With $p = \theta/a$, $\theta = \sqrt{\omega}$ we may write

$$X(x) = A \cosh px + B \sinh px + C \cos px + D \sin px.$$

The conditions $X(0) = X'(0) = 0$ readily imply that $C = -A$ and $D = -B$, so

$$
\begin{aligned}
X &= A(\cosh px - \cos px) + B(\sinh px - \sin px), \\
X' &= pA(\sinh px + \sin px) + pB(\cosh px - \cos px), \\
X'' &= p^2A(\cosh px + \cos px) + p^2B(\sinh px + \sin px), \\
X^{(3)} &= p^3A(\sinh px - \sin px) + p^3B(\cosh px + \cos px).
\end{aligned}
$$

The endpoint conditions at $x = L$ are

$$X''(L) = 0, \qquad (k - m\omega^2)X(L) - EIX^{(3)}(L) = 0.$$

When we substitute the derivatives above and write $z = pL$ we get

$$A(\cosh z + \cos z) + B(\sinh z + \sin z) = 0,$$

$$
\begin{aligned}
&A[(k - m\omega^2)(\cosh z - \cos z) - EIp^3(\sinh z - \sin z)] \\
+ &B[(k - m\omega^2)(\sinh z - \sin z) - EIp^3(\cosh z + \cos z)] = 0.
\end{aligned}
$$

If Δ denotes the coefficient determinant of these two linear equations in A and B, then the necessary condition $\Delta = 0$ for a non-trivial solution reduces eventually to the equation

$$
\begin{aligned}
EIp^3(1 &+ \cosh z \cos z) \\
&- (k - m\omega^2)(\sinh z \cos z - \cosh z \sin z) = 0.
\end{aligned}
$$

Finally we substitute $p = z/L$, $M = \rho L$, and

$$\omega^2 = p^4a^4 = (z^4/L^4)(EI/\rho)$$

to get the frequency equation

$$MEIz^3(1 + \cosh z \cos z)$$
$$= (kML^3 - mEIz^4)(\sinh z \cos z - \cosh z \sin z).$$

We may divide by $\cosh z \cos z$ to write it in the form

$$MEIz^3(1 + \operatorname{sech} z \sec z)$$
$$= (kML^3 - mEIz^4)(\tanh z - \tan z).$$

If β_n denotes the nth positive root of this equation, then the nth natural frequency is

$$\omega_n = (\beta_n/L)^2 \sqrt{(EI/\rho)}.$$

15. We want to calculate the fundamental frequency of transverse vibration of a cantilever with the numerical parameters

$$L = 400 \text{ cm}$$
$$E = 2 \cdot 10^{12} \text{ g/cm-sec}^2$$
$$I = (1/12)(30 \text{ cm})(2 \text{ cm})^3 = 20 \text{ cm}^4$$
$$\rho = (7.75 \text{ g/cm}^3)(60 \text{ cm}^2) = 465 \text{ g/cm}.$$

When we substitute these values and $\beta_1 = 1.8751$ in the frequency formula

$$\omega_1 = (\beta_1/L)^2 \sqrt{(EI/\rho)}$$

we find that $\omega_1 \approx 6.45$ rad/sec, so the fundamental frequency is $\omega_1/2\pi \approx 1.03$ cycles/sec. Thus the diver should bounce up and down on the end of the diving board about once every second.

16. When we substitute $y(x,t) = X(x)\cos \omega t$ in the given partial differential equation

$$\rho y_{tt} + P y_{xx} + EI y_{xxxx} = 0$$

and cancel the factor $\cos \omega t$, we get the ordinary differential equation

$$EIX^{(4)} + PX'' - \lambda X = 0$$

where $\lambda = \rho\omega^2$. By solving the characteristic equation

$$EIr^4 + Pr^2 - \lambda = 0$$

we find the general solution

$$X(x) = A \cosh \alpha x + B \sinh \alpha x + C \cos \beta x + D \sin \beta x$$

where

$$\alpha^2 = [- P + \sqrt{(P^2 + 4\lambda EI)}]/2EI,$$
$$\beta^2 = - [- P - \sqrt{(P^2 + 4\lambda EI)}]/2EI.$$

The endpoint conditions $X(0) = X''(0) = 0$ imply that $A = C = 0$, so

$$X(x) = B \sinh \alpha x + D \sin \beta x.$$

Then the conditions $X(L) = X''(L) = 0$ yield the equations

$$B \sinh \alpha L + D \sin \beta L = 0,$$
$$\alpha^2 B \sinh \alpha L - \beta^2 D \sin \beta L = 0.$$

The determinant of these two linear equations in B and D must vanish in order that a non-trivial solution exist, so

$$(\alpha^2 + \beta^2)\sinh \alpha L \sin \alpha L = 0.$$

It follows that $\sin \alpha L = 0$, so αL must be an integral multiple of π. The definitions of α^2 and β^2 imply that

$$\beta^2 - \alpha^2 = P/EI, \qquad \alpha^2\beta^2 = \lambda/EI.$$

Hence if $\alpha_n = n\pi/L$, the corresponding value of β_n is

$$\beta_n = (n^2\pi^2/L^2 + P/EI)^{1/2}.$$

Then the corresponding value of λ is

$$\lambda_n = EI(\alpha_n\beta_n)^2 = EI(n^4\pi^4/L^4)(1 + PL^2/n^2\pi^2EI).$$

Finally the nth natural frequency is

$$\omega_n = \sqrt{(\lambda_n/\rho)} = (n^2\pi^2/L^2)(1 + PL^2/n^2\pi^2EI)^{1/2}\sqrt{(EI/\rho)}.$$

19. We want to determine a steady periodic solution of the form

$$y(x,t) = X(x)\sin \omega t$$

where, with $\theta = \sqrt{\omega}$ and $p = \theta/a$,

$$X(x) = A \cosh px + B \sinh px + C \cos px + D \sin px$$

as usual. The fixed end conditions $X(0) = X'(0) = 0$ imply that $C = -A$ and $D = -B$, so

$$X(x) = A(\cosh px - \cos px) + B(\sinh px - \sin px).$$

It remains only to find A and B. But the endpoint conditions

$$X''(L) = EIX^{(3)}(L) + F_0 = 0$$

yield the linear equations

$$A(\cosh pL + \cos pL) + B(\sinh pL + \sin pL) = 0$$
$$A(\sinh pL - \sin pL) + B(\cosh pL + \cos pL) = -F_0/p^3EI$$

that can be solved for A and B.

22. When we substitute

$$e(x,t) = E(x)e^{i\omega t}$$

in the partial differential equation

$$e_{xx} = LCe_{tt} + (LG + RC)e_t + RGe$$

and cancel the factor $e^{i\omega t}$, the result is the ordinary differential equation

$$E''(x) - \gamma E(x) = 0$$

where

$$\gamma = (RG - LC\omega^2) + i\omega(LG + RC).$$

If $(\alpha + \beta i)^2 = \gamma$, then the general solution is

$$E(x) = Ae^{-\alpha x}e^{-i\beta x} + Be^{\alpha x}e^{i\beta x}.$$

In order that $e(x,t)$ be bounded as $x \longrightarrow \infty$ we choose $B = 0$, and in order that $e(0,t) = E_0\cos\omega t$ we choose $A = E_0$. Then our steady periodic solution is the real part

$$Re[E(x)e^{i\omega t}] = Re[E_0e^{-\alpha x}e^{-i\beta x}e^{i\omega t}]$$

$$= E_0e^{-\alpha x}\cos(\omega t - \beta x).\wedge 0$$

SECTION 9.4

APPLICATIONS OF BESSEL FUNCTIONS

2. The boundary value problem we want to solve is

$$u_{tt} = a^2[u_{rr} + (1/r)u_r] \qquad (r < c,\ t > 0)$$
$$u(c,t) = 0$$
$$u(r,0) = 0, \qquad u_t(r,0) = v_0$$

The substitution $u(r,t) = R(r)T(t)$ yields the equations

$$rR'' + R' + \alpha^2 rR = 0,$$
$$T'' + \alpha^2 a^2 T = 0$$

with separation constant $\lambda = \alpha^2$. The first is the parametric Bessel equation of order zero with general solution

$$R(r) = AJ_0(\alpha r) + BY_0(\alpha r).$$

In order that $R(r)$ be continuous at $r = 0$ we choose $B = 0$, so $R(r) = AJ_0(\alpha r)$. Then

$$R(c) = AJ_0(\alpha c) = 0$$

requires that αc be one of the roots $\{\gamma_n\}$ of $J_0(x) = 0$. With $\alpha_n = \gamma_n/c$ we get

$$R_n(r) = J_0(\gamma_n r/c).$$

The corresponding function $T(t)$ is

$$T_n(t) = A_n \cos \gamma_n at/c + B_n \sin \gamma_n at/c,$$

and we choose $A_n = 0$ to satisfy the condition $T(0) = 0$. Thus the displacement function $u(r,t)$ is of the form

$$u(r,t) = \sum c_n J_0(\gamma_n r/c) \sin \gamma_n at/c,$$

with

$$u_t(r,0) = \sum (c_n \gamma_n a/c) J_0(\gamma_n r/c).$$

In order to satisfy the initial condition $u_t(r,0) = v_0$ we apply Equation (22) and finally choose

$$c_n = \frac{c}{\gamma_n a} \cdot \frac{2}{c^2 J_1(\gamma_n)^2} \int_0^c r v_0 J_0(\gamma_n r/c) \, dr$$

$$= \frac{2cv_0}{a \gamma_n^3 J_1(\gamma_n)^2} \int_0^{\gamma_n} x J_0(x) \, dx$$

$$= \frac{2cv_0}{a \gamma_n^3 J_1(\gamma_n)^2} \left[x J_1(x) \right]_0^{\gamma_n} = \frac{2cv_0}{a \gamma_n^2 J_1(\gamma_n)}$$

3. (a) As in Problem 2

$$u(r,t) = \sum c_n J_0(\gamma_n r/c) \sin \gamma_n at/c.$$

In order to satisfy the given initial condition we must choose

$$c_n = \frac{2}{\gamma_n a \, c \, J_1(\gamma_n)^2} \int_0^{\epsilon} (P_0/\rho\pi\epsilon^2) \, r \, J_0(\epsilon_n r/c) \, dr$$

$$= \frac{2P_0 c}{\rho\pi\epsilon^2 \, \gamma^3 \, a \, J(\gamma_n)^2} \int_0^{\gamma_n \epsilon/c} x J_0(x) \, dx$$

$$= \frac{2P_0 c}{\rho\pi\epsilon^2 \gamma_n^3 a \, J_1(\gamma_n)^2} \cdot \frac{\gamma_n \epsilon}{c} J_1(\gamma_n \epsilon/c)$$

$$c_n = \frac{2aP_0}{\pi c\rho a^2 \gamma_n J_1(\gamma_n)^2} \cdot \frac{J_1(\gamma_n \epsilon/c)}{\gamma_n \epsilon/c}$$

(b) The final formula given in the text for $u(r,t)$ now follows because $\rho a^2 = T$ and $J_1(x)/x \longrightarrow 1$ as $x \longrightarrow 0$.

6. (a) We start with the boundary value problem

$$u_{rr} + (1/r)u_r + u_{zz} = 0 \qquad (r < c, \; 0 < z < L)$$
$$u_r(c,z) = 0$$
$$u(r,0) = 0, \qquad u(r,L) = f(r)$$

The substitution $u(r,z) = R(r)Z(z)$ yields the equations

$$rR'' + R' + \alpha^2 rR = 0,$$

$$Z'' - \alpha^2 Z = 0$$

with separation constant $\lambda = \alpha^2$. The homogeneous endpoint conditions are

$$R'(c) = Z(0) = 0.$$

If $\lambda = \alpha^2 = 0$ then $rR'' + R = 0$ implies

$$R(r) = A + B \ln r.$$

We choose $B = 0$ for continuity at $r = 0$, so $R(r) = A$. Then $R'(c) = 0$, so $\lambda_0 = 0$ is an eigenvalue, and we may take $R_0(r) = 1$. The equation $Z''(z) = 0$ implies $Z(z) = Az + B$, but $Z(0) = 0$ implies $B = 0$, so we take $Z_0(z) = z$.

If $\lambda = \alpha^2 > 0$ then we have the parametric Bessel equation with general solution

$$R(r) = AJ_0(\alpha r) + BY(\alpha r).$$

In order that $R(r)$ be continuous at $r = 0$ we choose $B = 0$, so $R(r) = AJ_0(\alpha r)$. Then

$$R'(c) = \alpha AJ_0'(\alpha c) = 0$$

requires that $\gamma = \alpha c$ be a root of

$$J_0'(x) = 0.$$

If $\alpha_n = \gamma_n/c$ where γ_n is the nth positive root, then

$$R_n(r) = J_0(\gamma_n r/c).$$

The corresponding function $Z(z)$ is

$$Z_n(z) = A_n \cosh \gamma_n z/c + B_n \sinh \gamma_n z/c,$$

and we choose $A_n = 0$ because $Z(0) = 0$. Thus we get the solution

$$u(r,z) = c_0 z + \sum c_n J_0(\gamma_n r/c) \sinh \gamma_n z/c$$

where $J_0'(\gamma_n) = 0$. To satisfy the condition $u(r,L) = f(r)$ we apply the formulas in (24) and choose

$$c_0 = \frac{2}{Lc^2} \int_0^c r\, f(r)\, dr$$

$$c_n = \frac{2}{c^2 \sinh(\gamma_n L/c) J_0(\gamma_n)^2} \int_0^c r\, f(r) J_0(\gamma_n r/c)\, dr$$

(b) If $f(r) = u_0$ (constant), then the coefficient formulas above readily yield $c_0 = u_0/L$ and $c_n = 0$ for $n > 0$, the latter because

$$\int x J_0(x)\, dx = x J_1(x) + C = -x J_0'(x) + C.$$

Hence the series reduces to the solution $u(r,z) = u_0 z/L$ that one might well guess without all this nonsense.

7. We want to solve the boundary value problem

$$u_{rr} + (1/r)u_r + u_{zz} = 0 \qquad (r < 1,\ z > 0)$$
$$h\, u(1,z) + u_r(1,z) = 0$$
$$u(r,z) \text{ bounded as } z \longrightarrow \infty$$
$$u(r,0) = u_0$$

We start with the separation of variables in Problem 6,

$$rR'' + R' + \alpha^2 rR = 0,$$

$$Z'' - \alpha^2 Z = 0$$

and readily see that $\alpha = 0$ is not an eigenvalue. When we impose the condition

$$hR(1) + R'(1) = 0$$

on $R(r) = J_0(\alpha r)$, we find that α must satisfy the equation

$$hJ_0(x) + xJ_0'(x) = 0$$

that corresponds to Case 2 with $n = 0$ in Figure 8.15. If $\{\gamma_n\}$ are the positive roots of this equation then

$$R_n(r) = J_0(\gamma_n r).$$

The general solution of $Z'' = (\gamma_n)^2 Z$ is

$$Z_n(z) = A_n \exp(-\gamma_n z) + B_n \exp(\gamma_n z),$$

and we choose $B_n = 0$ so that $Z_n(z)$ will be bounded as $z \longrightarrow \infty$. Thus we obtain a solution of the form

$$u(r,z) = \sum c_n \exp(-\gamma_n z) J_0(\gamma_n r)$$

where

$$hJ_0(\gamma_n) + \gamma_n J_0'(\gamma_n) = 0.$$

We readily calculate the coefficients to get the final formula given in the text.

11. When we substitute $u(r,t) = R(r)\sin \omega t$ in the given partial differential equation and cancel the factor $\sin \omega t$, we get the ordinary differential equation

$$R'' + (1/r)R' + (\omega^2/a^2)R = -F_0/\omega^2.$$

The solution that is continuous at $r = 0$ is

$$R(r) = AJ_0(\omega r/a) - F_0/\omega^2.$$

The condition $R(b) = 0$ yields

$$A = F_0/\omega^2 J_0(\omega b/a)$$

so it follows that

$$u(r,t) = [F_0/\omega^2 J_0(\omega b/a)][J_0(\omega r/a) - J_0(\omega b/a)]\sin \omega t.$$

12. When we substitute $y(x,t) = X(x)\sin \omega t$ in the partial differential equation

$$y_{tt} = (g/w)(wxy_x)_x = g(y_x + xy_{xx}),$$

we get the ordinary differential equation

$$x^2 X'' + xX' + (\omega^2 x/g)X = 0.$$

This is of the form of Equation (3) in Section 3.6 with $A = 1$, $B = 0$, $C = \omega^2/g$, and $q = 1$, so its general solution is

$$X(x) = AJ_0(2\omega\sqrt{(x/g)}) + BY_0(2\omega\sqrt{(x/g)}).$$

We choose $B = 0$ for continuity at $x = 0$, and the condition $X(L) = 0$ requires that $2\omega\sqrt{(L/g)}$ be one of the roots $\{\gamma_n\}$ of the equation $J_0(x) = 0$. Hence the nth natural frequency of vibration of the hanging cable is

$$\omega_n = (\gamma_n/2)\sqrt{(g/L)}.$$

13. With $w(x) = wx$ and $h(x) = h$ (w and h constant) the given partial differential equation reduces to

$$xy_{tt} = gh(y_x + y_{xx}).$$

When we substitute $y(x,t) = X(x)\cos \omega t$ we get the parametric Bessel equation

$$x^2 X'' + xX' + (\omega^2 x^2/gh)X = 0$$

with (finite) solution

$$X(x) = AJ_0(\omega x/\sqrt(gh)).$$

The condition $X = y_0$ implies that $A = y_0/J_0(\omega L/\sqrt(gh))$, so

$$y(x,t) = [y_0/J_0(\omega L/\sqrt(gh))]J_0(\omega x/\sqrt(gh))\cos \omega t.$$

14. With $w(x) = w$ and $h(x) = kx$ the given partial differential equation reduces to

$$y_{tt} = gh(y_x + xy_{xx}).$$

When we substitute $y(x,t) = X(x)\cos \omega t$ we get the ordinary differential equation

$$x^2 X'' + xX' + (\omega^2 x/gh)X = 0.$$

This has the form of Equation (3) in Section 3.6 with $A = 1$, $B = 0$, $C = \omega^2/gh$, and $q = 1$, so its (finite) solution is

$$X(x) = AJ_0(2\omega\sqrt(x/gh)).$$

The condition $X(L) = y_0$ now implies that

$$A = y_0/J_0(2\omega\sqrt(L/gh)).$$

16. With $\lambda = \alpha^2$ the general solution is

$$y(x) = AJ_0(\alpha x) + BY_0(\alpha x).$$

The endpoint conditions $y(a) = y(b) = 0$ yield the linear equations

$$AJ_0(\alpha a) + BY_0(\alpha a) = 0,$$
$$AJ_0(\alpha b) + BY_0(\alpha b) = 0.$$

In order for there to exist a non-trivial solution for A and B the coefficient determinant must vanish. Hence α must be one of the solutions $\{\gamma_n\}$ of the equation

$$J_0(ax)Y_0(bx) - J_0(bx)Y_0(ax) = 0.$$

With $\alpha = \gamma_n$, $A = Y_0(\gamma_n a)$ and $B = -J_0(\gamma_n a)$, both conditions above are satisfied and we have the eigenfunction

$$R_n(x) = Y_0(\gamma_n a)J_0(\gamma_n x) - J_0(\gamma_n a)Y_0(\gamma_n x).$$

18. We start with the substitution $u(r,t) = R(r)T(t)$ in the heat equation. The result is given in Equations (25) and (26):

$$r^2R'' + rR' + \alpha^2 r^2 R = 0,$$

$$T' = -\alpha^2 kT.$$

The first of these equations, together with the endpoint conditions
$$R(a) = R(b) = 0,$$

comprises the regular Sturm-Liouville problem of Problem 16. Hence its eigenvalues are given by $\alpha_n = \gamma_n$ where $\{\gamma_n\}$ are the positive roots of the equation in (41). The nth eigenfunction is the function $R_n(r)$ defined in (42). Finally the solution of $T_n' = -(\gamma_n)^2 kT_n$ is

$$T_n(t) = \exp(-(\gamma_n)^2 kt),$$

so we get a solution of the form

$$u(r,t) = \sum c_n \exp[-(\gamma_n)^2 kt]R_n(r).$$

NUCLEAR REACTORS AND OTHER APPLICATIONS

This section provides the interested student with an opportunity
to study a particular application at greater depth than is
afforded by the usual textbook exercises. For instance, each of
the "case studies" here could serve as the basis for a term paper
project. Because these problem sets are rather heavily annotated
in the text, further outlines of solutions are not included in
this manual.

EXISTENCE AND UNIQUENESS OF SOLUTIONS

In Problems 1-12 we apply the iterative formula

$$y_{n+1}(x) = b + \int_a^x f(t, y_n(t)) \, dt$$

to compute successive approximations $\{y_n(x)\}$ to the solution of the initial value problem

$$y' = f(x, y), \qquad y(a) = b.$$

starting with $y_0(x) = b$.

1. $y_0(x) = 3$

   $y_1(x) = 3 + 3x$

   $y_2(x) = 3 + 3x + 3x^2/2$

   $y_3(x) = 3 + 3x + 3x^2/2 + x^3/2$

   $y_4(x) = 3 + 3x + 3x^2/2 + x^3/2 + x^4/8$

   $y(x) \quad = 3e^x$

   $\qquad = 3 - 3x + 3x^2/2 + x^3/2 + x^4/8 + \cdots$

2. $y_0(x) = 4$

   $y_1(x) = 4 - 8x$

   $y_2(x) = 4 - 8x + 8x^2$

$$y_3(x) = 4 - 8x + 8x^2 - (16/3)x^3$$

$$y_4(x) = 4 - 8x + 8x^2 - (16/3)x^3 + (8/3)x^4$$

$$y(x) = 4e^{-2x}$$

$$= 4 - 8x + 8x^2 - (16/3)x^3 + (8/3)x^4 - \cdots$$

3. $y_0(x) = 1$

$$y_1(x) = 1 - x^2$$

$$y_2(x) = 1 - x^2 + x^4/2$$

$$y_3(x) = 1 - x^2 + x^4/2 - x^6/6$$

$$y_4(x) = 1 - x^2 + x^4/2 - x^6/6 + x^8/24$$

$$y(x) = \exp(-x^2)$$

$$= 1 - x^2 + x^4/2 - x^6/6 + x^8/24 - \cdots$$

4. $y_0(x) = 2$

$$y_1(x) = 2 + 2x^3$$

$$y_2(x) = 2 + 2x^3 + x^6$$

$$y_3(x) = 2 + 2x^3 + x^6 + (1/3)x^9$$

$$y_4(x) = 2 + 2x^3 + x^6 + (1/3)x^9 + (1/12)x^{12}$$

$$y(x) = 2\exp(x^3)$$

$$= 2 + 2x^3 + x^6 + (1/3)x^9 + (1/12)x^{12} + \cdots$$

5. $y_0(x) = 0$

$$y_1(x) = 2x$$

$$y_2(x) = 2x + 2x^2$$

$$y_3(x) = 2x + 2x^2 + 4x^3/3$$

$$y_4(x) = 2x + 2x^2 + 4x^3/3 + 2x^4/3$$

$$y(x) = e^{2x} - 1 = 2x + 2x^2 + 4x^3/3 + 2x^4/3 + \cdots$$

6. $y_0(x) = 0$

$y_1(x) = (1/2)x^2$

$y_2(x) = (1/2)x^2 + (1/6)x^3$

$y_3(x) = (1/2)x^2 + (1/6)x^3 + (1/24)x^4$

$y_4(x) = (1/2)x^2 + (1/6)x^3 + (1/24)x^4 + (1/120)x^5$

$y(x) = e^x - x - 1$

$\quad\quad = (1/2!)x^2 + (1/3!)x^3 + (1/4!)x^4 + (1/5!)x^5 + \cdots$

7. $y_0(x) = 0$

$y_1(x) = x^2$

$y_2(x) = x^2 + x^4/2$

$y_3(x) = x^2 + x^4/2 + x^6/6$

$y_4(x) = x^2 + x^4/2 + x^6/6 + x^8/24$

$y(x) = \exp(x^2) - 1$

$\quad\quad = x^2 + x^4/2 + x^6/6 + x^8/24 + \cdots$

8. $y_0(x) = 0$

$y_1(x) = 2x^4$

$y_2(x) = 2x^4 + (4/3)x^6$

$y_3(x) = 2x^4 + (4/3)x^6 + (2/3)x^8$

$y_4(x) = 2x^4 + (4/3)x^6 + (2/3)x^8 + (4/15)x^{10}$

$y(x) = \exp(2x^2) - 2x^2 - 1$

$\quad\quad = 2x^4 + (4/3)x^6 + (2/3)x^8 + (4/15)x^{10} + \cdots$

9. $y_0(x) = 1$

$y_1(x) = (1 + x) + x^2/2$

$$y_2(x) = (1 + x + x^2) + x^3/6$$

$$y_3(x) = (1 + x + x^2 + x^3/3) + x^4/24$$

$$y(x) = 2e^x - 1 - x$$

$$= 1 + x + x^2 + x^3/3 + x^4/12 + \cdots$$

10. $\quad y_0(x) = 0$

$\quad y_1(x) = e^x - 1$

$$= x + (1/2)x^2 + (1/6)x^3 + (1/24)x^4 + \cdots$$

$\quad y_2(x) = 2e^x - x - 2$

$$= x + x^2 + (1/3)x^3 + (1/12)x^4 + \cdots$$

$\quad y_3(x) = 3e^x - (1/2)x^2 - 2x - 3$

$$= x + x^2 + (1/2)x^3 + (1/8)x^4 + \cdots$$

$\quad y(x) = xe^x$

$$= x + x^2 + (1/2)x^3 + (1/6)x^4 + \cdots$$

11. $\quad y_0(x) = 1$

$\quad y_1(x) = 1 + x$

$\quad y_2(x) = (1 + x + x^2) + x^3/3$

$\quad y_3(x) = (1 + x + x^2 + x^3) + 2x^4/3 + x^5/3 + x^6/9 + x^7/63$

$\quad y(x) = 1/(1 - x)$

$$= 1 + x + x^2 + x^3 + x^4 + \cdots$$

12. $\quad y_0(x) = 1$

$\quad y_1(x) = 1 + (1/2)x$

$\quad y_2(x) = 1 + (1/2)x + (3/8)x^2 + (1/8)x^3 + (1/64)x^4$

$\quad y_3(x) = 1 + (1/2)x + (3/8)x^2 + (5/16)x^3 + (13/64)x^4 + \cdots$

$\quad y(x) = (1 - x)^{-1/2}$

$$= 1 + (1/2)x + (3/8)x^2 + (5/16)x^3 + (35/128)x^4 + \cdots$$

13.
$$\begin{bmatrix} x_0(t) \\ y_0(t) \end{bmatrix} = \begin{bmatrix} 1 \\ -1 \end{bmatrix}$$

$$\begin{bmatrix} x_1(t) \\ y_1(t) \end{bmatrix} = \begin{bmatrix} 1 + 3t \\ -1 + 5t \end{bmatrix}$$

$$\begin{bmatrix} x_2(t) \\ y_2(t) \end{bmatrix} = \begin{bmatrix} 1 + 3t + \frac{1}{2} t^2 \\ -1 + 5t - \frac{1}{2} t^2 \end{bmatrix}$$

$$\begin{bmatrix} x_3(t) \\ y_3(t) \end{bmatrix} = \begin{bmatrix} 1 + 3t + \frac{1}{2} t^2 + \frac{1}{3} t^3 \\ -1 + 5t - \frac{1}{2} t^2 + \frac{5}{6} t^3 \end{bmatrix}$$

14.
$$x(t) = \begin{bmatrix} \sum_0^\infty \frac{1}{n!} \begin{bmatrix} 1 & n \\ 0 & 1 \end{bmatrix} t^n \end{bmatrix} \begin{bmatrix} 1 \\ 1 \end{bmatrix}$$

$$= \begin{bmatrix} \sum t^n/n! & \sum t^n/(n-1)! \\ 0 & \sum t^n/n! \end{bmatrix} \begin{bmatrix} 1 \\ 1 \end{bmatrix}$$

$$= \begin{bmatrix} e^t & te^t \\ 0 & e^t \end{bmatrix} \begin{bmatrix} 1 \\ 1 \end{bmatrix}$$

$$x(t) = \begin{bmatrix} e^t + te^t \\ e^t \end{bmatrix}$$

16. $y_0(x) = 0$
$y_1(x) = (1/3)x^3$
$y_2(x) = (1/3)x^3 + (1/63)x^7$
$y_3(x) = (1/3)x^3 + (1/63)x^7 + (2/2079)x^{11} + (1/59535)x^{15}$

Then $y_3(1) \approx 0.350185$, which differs by only 0.0134% from the Runge-Kutta approximation $y(1) \approx 0.350232$. As a denouement we may recall from the result of Problem 16 in Section 3.6 that the exact solution of our initial value problem here is

$$y(x) = x \, J_{3/4}(x^2/2) / J_{-1/4}(x^2/2)$$

so the exact value at $x = 1$ is

$$y(1) = J_{3/4}(1/2) / J_{-1/4}(1/2).$$